Lecture Notes in Statistics 182

Edited by P. Bickel, P. Diggle, S. Fienberg, U. Gather,
I. Olkin, S. Zeger

T0074779

Lixing Zhu

Nonparametric Monte Carlo Tests
and Their Applications

With 15 Figures

 Springer

Lixing Zhu
Department of Statistics and Actuarial Science
University of Hong Kong and Chinese
    Academy of Sciences
Pokfulam Road
Hong Kong
lzhu@hku.hk

Library of Congress Control Number: 2005926894

ISBN-10: 0-387-25038-7
ISBN-13: 978-0387-25038-0   Printed on acid-free paper.

Camera ready copy provided by the editors.

Printed in the United States of America.   (EB)

9 8 7 6 5 4 3 2 1

springeronline.com

To my wife and son: Qiushi and Qingqing

Preface

Monte Carlo approximations to the distributions of statistics have become important tools in statistics. In statistical inference, Monte Carlo approximation is performed by comparing the distribution of a statistic based on the observed data and that based on reference data. How to generate reference data is a crucial question in this research area.

In a parametric setup, the Monte Carlo test (MCT) was first introduced by Barnard (1963), in his discussion of Bartlett's (1963) paper. MCT has some nice features and is very similar to the parametric bootstrap; see Beran and Ducharme (1991). There are several developments afterwards. The optimality and computational efficiency of MCT have been investigated. MCT has also been applied to approximating spatial distributions. To perform MCT, it is crucial that we can generate reference datasets which allow the test procedure to approximate the null distribution of the test. This is because we wish that the approximation procedure allow the test to maintain the significance level and to have good power performance. However, for distributions and models with semiparametric structures, generating such reference datasets is a challenge.

In these notes, we propose an alternative approach to attack the problems with semiparametric and nonparametric structures: that is, the nonparametric Monte Carlo test (NMCT). The NMCT is motivated by MCT and other Monte Carlo approximations such as the bootstrap for the distributions of test statistics. The algorithms of NMCT are easy to implement and, in some cases, the exact validity of the tests can hold. Furthermore, the accuracy of the approximations is also relatively easy to study. We will describe this in Chapter 3.

This book was written on the basis of my journal papers and the seminars at East China Normal University, Shanghai, China when I visited there as an adjunct chair Professor in summers. I am using the opportunity to thank my friends and students.

I owe much to my friends and colleagues Y. Fujikoshi, K. Naito, G. Neuhaus, K. W. Ng, W. Stute, and K. C. Yuen for their help and the stimulating conversations. The materials in this book are mostly due to the collaborations with them.

The first manuscript was read by my student Ms. W. L. Xu. Her careful reading helped me correct many typographical errors. Chapter 8 was jointly written by her and me because most of the material in this chapter will be a part of her PHD thesis. Chapter 6 is based on an unpublished paper jointly with R. Q. Zhu who was in charge of all simulations of this chapter. Dr. Z. Y. Zhu and Dr. Z. Q. Zhang also gave me some helpful comments when I gave seminars.

The work was done with the partial support of the University of Hong Kong and the Research Grants Council of Hong Kong, Hong Kong, China (#HKU7129/00P, #HKU7181/02H, #HKU7060/04P). As a Humboldt Research Award winner, I was also supported by the Alexander-von Humboldt Foundation of Germany for my academic visits to Giessen University and Hamburg University, Germany so that I had breaks during teaching to write this book.

I express my deep gratitude to John Kimmel for his constant help during the writing of this book, and my wife, son and my parents, especially my mother who passed away only a couple of months before the publication of this book, for their moral encouragement. Thanks also goes to Ada, who helped me solve many typing problems with Latex.

The University of Hong Kong, China
June, 2005   *Lixing Zhu*

Contents

1 Monte Carlo Tests 1
   1.1 Parametric Monte Carlo Tests 1
   1.2 Nonparametric Monte Carlo Tests (NMCT) 2
        1.2.1 The Motivation 2
        1.2.2 NMCT Based on Independent Decompositions 4
        1.2.3 NMCT Based on Random Weighting 6

2 Testing for Multivariate Distributions 11
   2.1 Four Classes of Multivariate Distributions 11
   2.2 A Test Statistic Based on Characteristic Function 12
   2.3 Simulations and Application 15
        2.3.1 Preamble 15
        2.3.2 Simulations 16
        2.3.3 Application 24

3 Asymptotics of Goodness-of-fit Tests for Symmetry 27
   3.1 Introduction 27
   3.2 Test Statistics and Asymptotics 28
        3.2.1 Testing for Elliptical Symmetry 28
        3.2.2 Testing for Reflection Symmetry 30
   3.3 NMCT Procedures 32
        3.3.1 NMCT for Elliptical Symmetry 32
        3.3.2 NMCT for Reflection Symmetry 35
        3.3.3 A Simulation Study 37
   3.4 Appendix: Proofs of Theorems 38

4 A Test of Dimension-Reduction Type for Regressions 45
   4.1 Introduction 45
   4.2 The Limit Behavior of Test Statistic 47
   4.3 Monte Carlo Approximations 48
   4.4 Simulation Study 50

       4.4.1 Power Study 50

       4.4.2 Residual Plots 51

       4.4.3 A Real Example. 52

  4.5 Concluding Remarks 53

  4.6 Proofs. 54

5 Checking the Adequacy of a Partially Linear Model 61

  5.1 Introduction 61

  5.2 A Test Statistic and Its Limiting Behavior 63

       5.2.1 Motivation and Construction 63

       5.2.2 Estimation of β and γ 64

       5.2.3 Asymptotic Properties of the Test 65

  5.3 The NMCT Approximation 66

  5.4 Simulation Study and Example 68

       5.4.1 Simulation Study 68

       5.4.2 An Example 70

  5.5 Appendix 72

       5.5.1 Assumptions 72

       5.5.2 Proof for Results in Section 5.2 73

       5.5.3 Proof for Results in Section 5.3 82

6 Model Checking for Multivariate Regression Models 85

  6.1 Introduction 85

  6.2 Test Statistics and their Asymptotic Behavior 86

       6.2.1 A Score Type Test 86

       6.2.2 Asymptotics and Power Study 87

       6.2.3 The Selection of $\boldsymbol{W}$ 89

       6.2.4 Likelihood Ratio Test for Regression Parameters 90

  6.3 NMCT Procedures 91

       6.3.1 The NMCT for $\boldsymbol{TT}_n$ 91

       6.3.2 The NMCT for the Wilks Lambda 93

  6.4 Simulations and Application 94

       6.4.1 Model Checking with the Score Type Test 94

       6.4.2 Diagnostics with the Wilks Lambda 96

       6.4.3 An Application 97

  6.5 Appendix 99

7 Heteroscedasticity Tests for Regressions 103

  7.1 Introduction 103

  7.2 Construction and Properties of Tests 104

       7.2.1 Construction 104

       7.2.2 The Limit Behavior of T_n and W_n 105

  7.3 Monte Carlo Approximations 107

  7.4 A Simulation Study 110

  7.5 Proofs of the theorems 114

7.5.1   A Set of Conditions...................................114
7.5.2   Proofs of the Theorems in Section 2 115
7.5.3   Proofs of the Theorems in Section 3 120

8   Checking the Adequacy of a Varying-Coefficients Model ..123
8.1   Introduction 123
8.2   Test Procedures 125
8.3   The Limit Behavior of Test Statistic 126
8.3.1   Innovation Process Approach 128
8.3.2   A Non-parametric Monte Carlo Test 130
8.4   Simulation Study and Application 131
8.4.1   Simulation Study.....................................131
8.4.2   Application to AIDS Data...........................133
8.5   Proofs..135

9   On the Mean Residual Life Regression Model 141
9.1   Introduction 141
9.2   Asymptotic Properties of the Test Statistic.................142
9.3   Monte Carlo Approximations...........................145
9.4   Simulations 147
9.5   Proofs 148

10   Homegeneity Testing for Covariance Matrices 155
10.1   Introduction 155
10.2   Construction of Tests 156
10.3   Monte Carlo Approximations 158
10.3.1 Classical Bootstrap 158
10.3.2 NMCT Approximation 159
10.3.3 Permutation Test...................................160
10.3.4 Simulation Study...................................161
10.4   Appendix 164

References 169

Index 179

1

Monte Carlo Tests

1.1 Parametric Monte Carlo Tests

Often in hypothesis testing the exact or limiting null distribution of a test statistic is intractable for determining critical values, leading to the use of Monte Carlo approximations. As an easily implemented methodology, the Monte Carlo test (MCT) has received attention in the literature. In his discussion of Bartlett's paper (1963), Barnard (1963) first described the idea of MCT. Hope (1968) proved that in the parametric context, if there are no nuisance parameters Monte Carlo tests admit exact significance levels and have very high power even when compared with uniformly most powerful (UMP) tests. MCT is also applicable when there are nuisance parameters; that is, MCT can be efficiently applied in parametric settings. Besag and Diggle (1977) used MCT in spatial patterns when there are nuisance parameters in the distribution of random variables. Engen and Lillegård (1997) applied MCT when the necessary simulations can be conditioned on the observed values of a minimal sufficient statistic under the null hypothesis. In certain circumstances with nuisance parameters, Monte Carlo tests still admit the exact levels of significance. Zhu, Fang, and Bhatti (1997) constructed a projection pursuit type Crämer-von Mises statistic to the testing for distribution in a parametric class. Hall and Titterington (1989) showed that in parametric settings with or without nuisance parameters, the level error of a MCT is an order of magnitude less than that of the corresponding asymptotic test whenever the test statistic is asymptotically pivotal, and a MCT is able to distinguish between the null hypothesis and alternative distinct $n^{-1/2}$ from the null. These results reinforced the use of MCT.

We use a simple example to explain the use of MCT. Consider independent identically distributed ($i.i.d.$) random variables $\{x_1, \cdots, x_n\}$ having a distribution $F(\cdot)$. Suppose that we want to test whether $F(\cdot) = G(\cdot, \theta)$ for some unknown parameter θ where $G(\cdot)$ is a given function. In this circumstance, for any test statistic of choice, say, $T(x_1, \cdots, x_n)$, a MCT requires the construction of a reference set of values of the test statistic by sampling from

$T(x'_1, \cdots, x'_n)$ where $x'_1, \cdots, x'_n$ are independently drawn from the distribution $G(\cdot, \hat{\theta})$ with $\hat{\theta}$ being an estimator of θ. Suppose that the null hypothesis will be rejected for large values of T; for two-sided tests, modifications are easily done. Let $T(x_1, \cdots, x_n) =: T_0$ and $T_1, \cdots, T_m$ be obtained from the Monte Carlo procedure. The p-value is estimated as

$$\hat{p} = k/(m+1),$$

where k is the number of values in $T_0, T_1, \ldots, T_m$ that are larger than or equal to T_0. Therefore, for given nominal level α, whenever $\hat{p} \leq \alpha$, the null hypothesis will be rejected.

It is worthwhile to point out that parametric bootstrap approximations developed in 1980's are similar to the above procedure, see, *e.g.* Beran and Ducharme (1991).

1.2 Nonparametric Monte Carlo Tests (NMCT)

1.2.1 The Motivation

In semiparametric or nonparametric settings, however, we may have difficulty in simulating reference datasets under the null hypothesis in order to perform a MCT. The major difficulty is that even under the null hypothesis, the model cannot be expressed with a specific structure up to a finite number of unknown parameters. For example, consider testing whether the distribution of the collected data is in the class of elliptically symmetric distributions (elliptical distribution for short). A d-dimensional random vector X is said to have an elliptical distribution if for any $d \times d$ orthogonal matrix H, there are a shape matrix A and a location μ such that the distribution of $HA(X - \mu)$ is identical with that of $A(X - \mu)$. When the second moment, Σ, of X is finite, A is $\Sigma^{-1/2}$. See Fang, Kotz, and Ng (1990) for details. From this definition, we know that the class of elliptical distributions is too large to be a parametric class. One of the powerful methods to use is the *Bootstrap*. Efron (1979) proposed this now time-honored methodology, and now it has become one of the most commonly used solutions to the above difficulty. The basic idea of Efron's bootstrap, which will be called Classical Bootstrap later, is to draw the reference datasets from the empirical distribution of the collected data. There are a number of variants and developments in the literature. See e.g., Davison and Hinkley (1997) for details. Shao and Tu (1995) is also a comprehensive reference. However, several concerns need to be tackled. First, the accuracy of the approximations is difficult to study. The research of accuracy or asymptotic accuracy is still from case to case, there is no a unified approach and not many papers deal with this issue. Related work is, for instance, Singh (1981). In the one-dimensional case, Zhu and Fang (1994) derived the accurate distribution of the bootstrap counterpart of the Kolmogorov statistic and obtained the $\sqrt{n}$ consistency. To our knowledge, this is the only paper dealing with the accurate distribution of bootstrap statistics. Second, the bootstrap approximations cannot make a test exactly valid, but possibly asymptotically valid because the reference data are from the empirical distribution that converges in distribution to the underlying distribution of the collected data; Third, bootstrap approximations are sometimes inconsistent and the correction for inconsistency is still from case to case. The m out of n bootstrap is a solution for correcting for the inconsistency, but in many cases, the m out of n bootstrap is less efficient. A variant proposed by Wu (1986) for reducing the bias of variance estimation in regression analysis and well developed by Mammen (1992), who named it *Wild Bootstrap* , is a powerful alternative solution. Wild bootstrap is successfully used in several areas, especially model checking for regression models; see Härdle and Mammen (1993) and Stute, González Manteiga, and Presedo Quindimil (1998). In some cases, it can overcome the inconsistency of Efron's classical bootstrap approximation. However, it fails to work in some other cases. Chapter 4 shows an example of its inconsistency when we consider the dimension-reduction type test for regression functions. Chapter 6 also provides a similar example when testing for heteroscedasticity. Fourth, in hypothesis testing, the bootstrapping sample needs to be delicate so as not to deteriorate the power of the tests.

The permutation test is another option. See Good (2000). It can be exactly valid in some cases. However, in one-sample cases, its applications are restrictive because most of the time we have difficulty in permuting the data so as to obtain the reference data. Its implementation is also time consuming.

Note that the bootstrap is a fully nonparametric methodology which needs few conditions on the model structure and the underlying distribution of data. Therefore, when the model is not fully nonparametric, but semi-structured, say the class of elliptical distributions, we may use other Monte Carlo approximations that can better use the information provided by the data. Based on these observations, we propose Nonparametric Monte Carlo Test (NMCT) Procedures. In Chapter 2, we will apply the NMCT to the testing problems for four types of distributions and will show its exact validity. In Chapter 3, we prove that when there are some nuisance parameters in the distributions studied in Chapter 2, the NMCT can obtain the asymptotic validity with the $\sqrt{n}$ consistency. This is what bootstrap approximation cannot achieve. In Chapters 4–6, we consider the model checking for regression models. The inconsistency of Wild bootstrap will be shown in Chapters 4 and 6. In Chapters 7–9, we consider some other problems in which bootstrap approximations work. But the NMCT procedures are easier to implement and the power performance is better. In the following two subsections, we separately describe the NMCT procedures when the random variables are independently decomposable and when the test statistics are asymptotically the functionals of linear statistics.

1.2.2 NMCT Based on Independent Decompositions

We motivate our procedures from the testing for several kinds of multivariate distributions that have played important roles and then propose a generic method. The detailed investigation on testing for multivariate distributions is presented in Chapter 2.

There are four commonly used classes of multivariate distributions. They are elliptically symmetric, reflectively symmetric, Liouville-Dirichlet and symmetric scale mixture distributions. The definitions of these classes of distributions will be presented in Chapters 2 and 3. These distributions are respectively the generalizations of normal, symmetric, Beta and stable distributions. See Fang, Kotz and Ng (1990), and references therein.

Tests of the adequacy of elliptically symmetric and reflectively symmetric distributions have received much attention; see, for example, Aki (1993), Baringhaus (1991), Beran (1979), Ghosh and Ruymgaart (1992), Heathcote, Rachev, and Cheng (1995). Note that the classes, for example, of elliptically symmetric distributions, are nonparametric, that is, the classes cannot be characterized by finite parameters. Therefore, we cannot simply apply a Monte Carlo test mentioned in Section 1.1 to approximate the sampling null distribution of a test statistic because the distribution of data is intractable. Also often in hypothesis testing the limiting null distribution of a test statistic is intractable for determining critical values. See, for instance, Zhu, Fang, Bhatti, and Bentler (1995). Diks and Tong (1999) proposed a conditional Monte Carlo test. When a density function is invariant under a compact group G of isometrics, the set of G-orbits is a minimal sufficient statistic and the simulation can be conditioned on the observed values of the G-orbits. They applied the method to testing for spherical and reflection symmetries of multivariate distributions without nuisance parameters. Neuhaus and Zhu (1998) and Zhu and Neuhaus (2003) also constructed conditional test procedures for these two types of symmetry of multivariate distributions.

We introduce the following method of generating reference datasets. The method relies on a property of independent decomposition of distribution.

DEFINITION 1.2.1 *A random vector X is said to be independent decomposable if $X = Y \bullet Z$ in distribution, Y and Z are independent and $Y \bullet Z$ is a dot product, that is $Y \bullet Z = (Y^{(1)}Z^{(1)}, \ldots, Y^{(d)}Z^{(d)})$ if both Y and Z are d-dimensional vectors, and $Y \bullet Z = (Y^{(1)}Z, \ldots, Y^{(d)}Z)$ if Z is scalar, where the $Y^{(i)}$'s are the components of Y; similarly, $Y \bullet Z = (YZ^{(1)}, \ldots, YZ^{(d)})$ if Y is scalar.*

When either Y or Z has an analytically tractable distribution, the above decomposition motivates the following MCT procedure. Let $x_1, \ldots, x_n$ be an i.i.d. sample of size n. A test statistic $T(x_1, \ldots, x_n)$, say, can be rewritten as $T(y_1 \bullet z_1, \ldots, y_n \bullet z_n)$, if the x_i's are independently decomposable with $x_i = y_i \bullet z_i$ under the null hypothesis. NMCT requires the construction of a reference set of values of the test statistic by sampling from $T(y'_1 \bullet z_1, \ldots, y'_n \bullet z_n)$, where $y'_1, \ldots, y'_n$ have the same distribution as that of $y_1, \ldots, y_n$. That is, conditionally on $z_1, \ldots, z_n$, the Monte Carlo calculations can be performed based on $y'_1, \ldots, y'_n$, where $y'_1, \ldots, y'_n$ are drawn independently from the distribution of Y. The p-value of the test statistic T can be estimated as follows. Suppose that the null hypothesis will be rejected for large values of T; for two-sided tests, modifications are easily done. Let the values of T be T_0 for the original data set and $T_1, \ldots, T_m$ obtained from the Monte Carlo procedure. The p-value is estimated as

$$\hat{p} = k/(m+1),$$

where k is the number of values in $T_0, T_1, \ldots, T_m$ that are larger than or equal to T_0. Therefore, for given nominal level α, whenever $\hat{p} \leq \alpha$, the null hypothesis will be rejected.

Since $T(x_1, \ldots, x_n)$ and $T(y'_1 \cdot z_1, \ldots, y'_n \cdot z_n)$ have the same distribution and also have the same conditional distribution given $z_1, \ldots, z_n$, exact validity of the test can be expected. The following proposition states this property.

PROPOSITION 1.2.1 *Under the null hypothesis that the vector X is independently decomposable $Y \cdot Z$, then for any $0 < \alpha < 1$*

$$Pr(\hat{p} \leq \alpha) \leq \frac{[\alpha(m+1)]}{m+1},$$

where the notation $[c]$ stands for the integer part of c.

Justification. Under the null hypothesis, $T_0, \ldots, T_m$ are independent identically distributed given $z_1, \ldots, z_n$, and $\hat{p} \leq \alpha$ implies $k \leq [\alpha(m+1)]$. In the case where there is no tie among the T_i's, $\hat{p}$ is distributed uniformly on the set $\left\{ \frac{1}{m+1}, \ldots, \frac{m+1}{m+1} \right\}$, so that

$$Pr(\hat{p} \leq \alpha \mid z_1, \ldots, z_n) = \frac{[\alpha(m+1)]}{m+1}.$$

If there are ties, $k \leq [\alpha(m+1)]$ and the strict inequality may hold. Then T_0 is larger than at least $m + 1 - [\alpha(m+1)]$ of T_i's. Hence

$$Pr(\hat{p} \leq \alpha \mid z_1, \ldots, z_n) \leq \frac{[\alpha(m+1)]}{m+1}.$$

The proof is concluded by integrating out over the z_i's.   □

This proposition clearly shows the exact validity of the NMCT when the variable is independently decomposable. In contrast, the bootstrap and permutation tests do not have such an advantage. In Chapter 2 we will apply it to the testing for the aforementioned four types of distributions.

1.2.3 NMCT Based on Random Weighting

When the hypothetical distribution of data does not have the property of independent decomposition, we suggest in this subsection a random weighting approach to generate reference datasets. This approach is rooted in empirical process theory: the random weighting method for convergence of stochastic processes.

Suppose that $\{x_1, \cdots, x_n\}$ is an $i.i.d.$ sample. Consider a test statistic, say $T_n = T(x_1, \cdots, x_n)$. When T_n can be rewritten as $\mathcal{T} \circ R_n$ where $\mathcal{T}$ is a functional on R_n, R_n is a process with the form

$$R_n = \{\frac{1}{\sqrt{n}} \sum_{j=1}^{n} J(x_j, t), \ t \in S\}$$

for a subset $S \in R^d$ and an integer d, or a random variable if S is a single point set where $E(J(X, t)) = 0$. We can use $T_n(E_n) = \mathcal{T} \circ R_n(E_n)$ as the conditional counterpart of T_n to construct an approximation of the distribution of T_n where

$$R_n(E_n) = \{\frac{1}{\sqrt{n}} \sum_{j=1}^{n} e_j J(x_j, t), \ t \in S\}$$

and $E_n = \{e_1, \cdots, e_n\}$ are random variables independent of Z_i. When e_j's are equally likely ± 1, this random weighting approach is called random symmetrization (Pollard, 1984), and when e_j's are normally distributed with mean 0 and variance 1, it has been used by Dudley (1978) and Giné and Zinn (1984). This algorithm is similar to Wild bootstrap (see, Mammen (1992)). When $\{x_1, \cdots, x_n\}$ are exchangeable random variables, van der Vaart and Wellner (2000) formulated this algorithm as Exchangeable Bootstrap.

However, most of test statistics cannot have such a simple representation. More often, a test statistic $T_n(x_1, \cdots, x_n)$ can be expressed asymptotically as follows:

$$T_n(x_1, \cdots, x_n, P_n) = \mathcal{T} \circ R_n + o_p(1), \qquad (1.2.1)$$

where R_n is a process with the form $\frac{1}{\sqrt{n}} \sum_{j=1}^{n} J(x_j, \psi, t)$ with $E(J(X, \psi, t)) = 0$, ψ is an unknown parameter of interest and ψ may be infinite dimensional, say, an unknown smooth function. Therefore a generic procedure is as follows:

- **Step 1**. Generate random variables $e_j, j = 1, ..., n$ independent with mean zero and variance one. Let $E_n := (e_1, \cdots, e_n)$ and define the conditional counterpart of G_n as

$$R_n(E_n, t) = \frac{1}{\sqrt{n}} \sum_{j=1}^{n} e_j J(x_j, \hat{\psi}, t), \qquad (1.2.2)$$

where $\hat{\psi}$ is a consistent estimator of ψ based on $\{x_1, \cdots, x_n\}$. The resultant conditional test statistic is

$$T_n(E_n) = \mathcal{T} \circ R_n(E_n). \tag{1.2.3}$$

- **Step 2**. Generate m sets of E_n, say $E_n^{(i)}, i = 1, ., m$, and then get m values of $T_n(E_n)$, say $T_n(E_n^{(i)}), i = 1, ..., m$.
- **Step 3**. Assume that we reject the null hypothesis for large value of T_n. The modification is easily done for two-sided tests. The p-value is estimated by $\hat{p} = k/(m+1)$ where k is the number of $T_n(E_n^{(i)})$'s which are larger than or equal to T_n. Reject H_0 when $\hat{p} \le \alpha$ for a designed level α.

PROPOSITION 1.2.2 *Assume that e_i's are i.i.d. and compactly supported, R_n converges in distribution to a continuous Gaussian process R, and $\hat{\psi} - \psi = O_p(n^{-a})$ for some $a > 0$. Furthermore, assume that for any fixed $t \in S$ the function J has second partial derivatives with respect to ψ, and all partial derivatives have finite first moments uniformly over all t. Then for almost all sequences $(x_1, \cdots, x_n)$, $T_n(E_n)$ has the same limit as T_n.*

Justification.   By the designed conditions, we can derive that

$$R_n(E_n, t) = \frac{1}{\sqrt{n}} \sum_{j=1}^{n} e_j J(z_j, \psi, t) + o_p(1). \tag{1.2.4}$$

Let $\mathcal{J} = \{J(\cdot, t) : t \in S\}$. In other words, $R_n(E_n, \cdot)$ is an empirical process indexed by a class of functions. Invoking Theorem 3.6.13 of van der Vaart and Wellner (2000), $R_n(E_n)$ has the same limit as R_n. The proof is concluded from the continuity of $\mathcal{T}$. $\qquad\square$

REMARK 1.2.1 *The following examples show some tests that can be expressed asymptotically as functionals of linear statistics in (1.2.1). Consider a regression model*

$$Y = \Phi(X) + \varepsilon$$

where $\Phi(\cdot)$ is an unknown function, Y is 1-dimensional response random variable, and X is p-dimensional column random vector independent of random variable ε. When we want to test the hypothesis

$$H_0: \quad \Phi(\cdot) \in \{\Phi_0(\cdot, \theta) : \theta \in \Theta\},$$

where Φ_0 is a given function and Θ is a compact set of q-dimensional Euclidean space R^q. Thus, we can find a column vector θ_0 such that $\Phi(\cdot) = \Phi_0(\cdot, \theta_0)$ under the null hypothesis. A commonly used method is to use residuals to construct a test statistic. When residuals are more likely to be away from zero, the values of the test statistic are large and the null hypothesis is rejected. From this idea, the following test statistics can be considered.

Suppose that $\{(x_1, y_1), \cdots, (x_n, y_n)\}$ is an *i.i.d.* sample.

A score type test. Let $\hat{\varepsilon}_j = y_j - \hat{\Phi}_0(x_j, \hat{\theta}_0)$, $j = 1, \cdots, n$ be the residuals obtained by fitting the hypothetical regression function $\Phi_0(\cdot, \theta_0)$ where $\hat{\theta}_0$ is an estimator of θ_0. A score type test can be defined by

$$T_n = [\frac{1}{\sqrt{n}} \sum_{j=1}^{n} \hat{\varepsilon}_j w(x_j, \hat{\theta}_0)]^2$$

where $w(\cdot)$ is a weight function to be selected and $\hat{\theta}_0$ is a consistent estimator. Let $R_n = \frac{1}{\sqrt{n}} \sum_{j=1}^{n} \hat{\varepsilon}_j w(x_j, \hat{\theta}_0)$. When $\hat{\theta}_0$ is a least squares estimator, it is easy to see that under H_0 and certain regularity conditions, $\hat{\theta}_0 - \theta_0$ can have an asymptotic linear representation with the form $\hat{\theta}_0 - \theta_0 = \frac{1}{n} \sum_{j=1}^{n} J_1(x_j, y_j, E(\Phi_0')^2, \theta_0) + o_p(1/\sqrt{n})$ where

$$J_1(x_j, y_j, E(\Phi_0')^2, \theta_0) =: [E(\Phi_0'(X, \theta_0))(\Phi_0'(X, \theta_0))^\tau]^{-1} \Phi_0'(x_j, \theta_0)\varepsilon_j$$

and Φ_0' is the first derivative of Φ_0 with respect to θ. The notation "τ" stands for transpose. Clearly, the expectation $E(J_1(X, Y, E(\Phi_0')^2, \theta_0)) = 0$. Furthermore, let

$$J(x_j, y_j, E(\Phi_0')^2, E(\Phi_0'w), \theta_0)$$
$$= \varepsilon_j w(x_j, \theta_0) - (J_1(x_j, y_j, E(\Phi_0')^2, \theta_0))^\tau E[\Phi_0'(X, \theta_0)w(X, \theta_0)].$$

We then obtain that asymptotically

$$R_n(t) = 1/\sqrt{n} \sum_{j=1}^{n} J(x_j, y_j, E(\Phi_0')^2, E(\Phi_0'w), \theta_0).$$

The functional $\mathcal{T}$ in (1.2.1) is a quandratic functional.

Cramér-von Mises type and Kolmogorov type tests. Let $R_n(x) = \frac{1}{\sqrt{n}} \sum_{j=1}^{n} \hat{\varepsilon}_j w(x_j, \theta_0) I(x_j \leq x)$ where "$X \leq x$" means that each component of X is less than or equal to the corresponding component of x. Similar to the score type test, we can also write R_n, asymptotically, as a linear sum $\frac{1}{\sqrt{n}} \sum_{j=1}^{n} J(x_j, y_j, E(\Phi_0')^2, E(\Phi_0'w), \theta_0, w, x)$ with a function $J(\cdot)$. In this case, the Cramér-von Mises type test statistic is $T_n = \int [R_n(X)]^2 dF(X)$ and the Kolmogonov type test statistic is $\sup_{t,x} |R_n(x)|$. The $\mathcal{T}$ will be the integration and supremum functionals respectively.

It should be pointed out that the algorithms developed here are similar to Wild bootstrap (see, e.g. Härdle and Mammen (1993) and Stute, González Manteiga and Presedo Quindimil (1998)). The difference is that Wild bootstrap draws the sample (X_i^*, Y_i^*)'s when the random variables e_i's multiply the residuals $\hat{\varepsilon}_i$, while NMCT places e_i's on the summands of R_n. We can prove that when testing the adequacy of linear models, that is, $\Phi_0(x, \theta_0) = \theta_0^\tau x$, Wild bootstrap is equivalent to NMCT if we use the above tests. The equivalence does not necessarily hold when other tests are used. In Chapter 4, we will provide mode details. For more general models, when the Crämer-von Mises test is used, the equivalency between NMCT and the Wild bootstrap also fails. We will discuss this in Chapter 5.

2

Testing for Multivariate Distributions

In this chapter, we investigate testing for multivariate distribution. For multivariate analysis, although testing for multivariate normality is still a topic, much effort has been devoted to nonparametric settings. In multivariate distribution theory, there are several important classes of distributions. In this chapter, we consider four kinds of distributions. Most of material in this chapter comes from Zhu and Neuhaus (2000).

In order to apply the methodology of NMCT developed in Chapter 1 to testing for multivariate distributions, we need to analyze which distributions have the property of independent decomposition. Let X be a d-dimensional random variable. Write it as $X = Y \cdot Z$ in distribution. We study in which cases Y and Z are independent and the distribution of Y is analytically tractable.

2.1 Four Classes of Multivariate Distributions

- Case (a). Spherically symmetric distributions.
  Here $X = U \bullet \|X\|$ in distribution, where U is independent of $\|X\|$ and is distributed uniformly on the sphere $S^d = \{a : \|a\| = 1, a \in R^d\}$ and $\|\cdot\|$ is the Euclidean norm in R^d. Let $Y = U$ and $Z = \|X\|$. The multivariate t distribution (Fang, Kotz, and Ng, 1990, Example 2.5) and the normal distribution $N(0, I_d)$ belongs to this class. In practical use, $U = X/\|X\|$.
- Case (b). Reflectively symmetric distributions.
  Here $X = -X$ in distribution, where X has the independent decomposition as $X = e \bullet X$ in distribution, in which $e = \pm 1$ with probability one half. Let $Y = e$ and $Z = X$. The uniform distribution on the cube $[-c, c]^d$, for positive c, is a member of this class.
- Case (c). Liouville-Dirichlet distributions.
  Here $X = Y \bullet r$ in distribution, where Y is independent of r, a scalar variable, and has a Dirichlet distribution $D(\alpha)$ with known parameter $\alpha = (\alpha_1, \cdots, \alpha_d)$ on the simplex $B^d = \{(y^{(1)}, \ldots, y^{(d)}) \in R^d : y^{(i)} \geq 0, \Sigma_{i=1}^d y^{(i)} = 1\}$. Let $Y = X/(\sum_{i=1}^d x^{(i)})$ and $r = \sum_{i=1}^d x^{(i)}$ where $X =$

$(x^{(1)}, \cdots, x^{(d)})$. The class includes multivariate Beta and inverted Dirichlet distributions (Olkin and Rubin (1964); Guttman and Tiao (1965)).

- Case (d). Symmetric scale mixture distributions.
  Here there is a scalar function $g(x)$ satisfying $g(x) = g(-x)$ and $g(x) \neq 0$ for $x \neq 0$ such that $x/g(x)$ is independent of $g(x)$ and is distributed uniformly on the space $C^d = \{y = (y^{(1)}, \ldots, y^{(d)}) \in R^d : g(y) = 1\}$. Let $Y = X/g(X)$ and $Z = g(X)$. This is a broad class of distributions including all spherically symmetric distributions. The multivariate extension of the Laplace distribution with density function $c \exp(-\sum_{i=1}^{d} |x^{(i)}|)$ is one of its members.

For the above types of distribution, one can generate y'_i from the uniform distribution on S^d for Case (a), the two-point distribution for Case (b), the Dirichlet distribution on B^d for Case (c) and the uniform distribution on C^d for Case (d).

The above classes of distributions do not contain nuisance parameters. For the testing problems in Section 2.3, we consider the distributions with nuisance location parameters. Let μ be the location of the distribution of X such as the mean or median.

When, in distribution,

$$X - \mu = Y \bullet Z,$$

often Z can be expressed as a function of $(X - \mu)$, $h(x - \mu)$, say. Consider the following distributions that are associated with Cases (a), (b) and (d):

- Case (a1). Spherical symmetry about μ.

$$X - \mu = U \bullet \|X - \mu\| \qquad \text{in distribution}$$

- Case (b1). Reflection symmetry about μ.

$$X - \mu = -(X - \mu) \qquad \text{in distribution}$$

- Case (d1). Symmetric scale mixture with unknown location μ.

Here $(x - \mu)/g(x - \mu)$ is independent of $g(x - \mu)$ and $\frac{x-\mu}{g(x-\mu)}$ is distributed uniformly on the space $C^d = \{y = (y_1, \ldots, y_d) \in R^d, g(y) = 1\}$.

For these three cases, $h(x - \mu)$ will be, respectively, $\|x - \mu\|$, $(x - \mu)$ and $g(x - \mu)$.

2.2 A Test Statistic Based on Characteristic Function

Let $\varphi_x(t) = E\{\exp(it'X)\}$ be the characteristic function of a multivariate distribution. When $X = Y \bullet Z$ in distribution, $\varphi_x(t)$ can be rewritten as

$\varphi_x(t) = E_Z\{\varphi_Y(t'Z)\}$ where φ_Y is the characteristic function of Y. Consider the integral

$$\int \left\| \varphi_x(t) - E_z\{\varphi_Y(t'z)\} \right\|^2 W_a(t)\, dt,$$

where $W_a(\cdot)$ is a continuous weight function, with parameter a, on its support set such that the integral is finite. Different but related statistics were used in goodness-of-fit tests for multivariate normality by Henze and Wagner (1997). The integral equals zero if the null hypothesis is true. Under the alternative, the positivity of the integral may be related in a certain manner to the choice of the weight function. If the support of $W_a(.)$ is R^d, as with a normal density function, the integral will be positive. When the support is a compact subset of R^d, this may not hold, so that a test based on the integral may not be consistent against all fixed alternatives. However, the problem is not very serious as the following example shows. Assume that the support of W_a is compact and contains the origin of R^d, that the characteristic functions are continuous and that the underlying distribution of X has moments of all orders. It can be shown as follows that the integral equals zero if and only if the two characteristic functions are equal. For the integral to be zero, the two characteristic functions must be equal almost everywhere and then everywhere in the support set by continuity. Since the support set includes the origin, Taylor expansion of the characteristic functions at the origin implies equality of all moments of the two distributions. Thus, the two distributions are the same. The sufficiency is clear.

The test statistic is constructed by replacing the underlying distribution by the empirical distribution. The test is consistent against any fixed alternative when the weight function is chosen suitably. In this chapter, we consider the normal and uniform density functions.

Note that the term $E_z\{\varphi_Y(t'z)\}$ involves the integral of Y. This may not have a simple analytic form. Because the distribution of Y is analytically tractable, we could approximate it by NMCT.

Under the null hypothesis, X is independently decomposable, we have $\int \left[\varphi_x(t)\overline{E_z}\{\varphi_Y(t'z)\}\right] W_a(t)\, dt = \int \left\| E_z\{\varphi_Y(t'z)\} \right\|^2 W_a(t)\, dt$ where $\overline{f}(t)$ denotes the dual function of $f(t)$. Hence,

$$\int \left\| \varphi_x(t) \right\|^2 W_a(t)\, dt - \int \left\| E_z\{\varphi_Y(t'z)\} \right\|^2 W_a(t)\, dt = T - T_1,$$

say. An estimator of T is its empirical counterpart: if F_{nx} is the empirical distribution of $x_1, \cdots, x_n$, then

$$T_n = \int \left\| \int e^{it'x} dF_{nx}(x) \right\|^2 W_a(t)\, dt$$

$$= \left\{ \int \left| \int \cos(t'x) dF_{nx}(x) \right|^2 W_a(t)\, dt + \int \left| \int \sin(t'x) dF_{nx}(x) \right|^2 W_a(t)\, dt \right\}$$

is clearly a consistent estimator of T when the distribution of X, F_x, is continuous.

Analogously, T_1, if F_{nz} is the empirical distribution of $z_1, \ldots, z_n$, can be estimated by

$$T_{n_1} = \int \left\| \int \varphi_Y(t'z)dF_{nz} \right\|^2 W_a(t)\, dt \ .$$

As mentioned above, the integral of Y needs not have a simple analytic form, but the evaluation of the integral is not necessary. Whenever $z_1, \ldots, z_n$ are given, T_{n_1} is a constant under the test procedures proposed in Section 1.2 of Chapter 1. Omitting from the test statistic this constant term and the values of its conditional counterpart generated by NMCT will have no effect on the p-values determined. Hence we need focus only on T_n in NMCT procedure.

The integral with respect to t takes closed forms with some commonly used weights such as normal density weight and uniform density weight.

PROPOSITION 2.2.1 Let $W_a(t) = (2\pi a)^{-d/2} \exp\left(-\|t\|^2/2a^2\right)$

$$T_n = \frac{1}{n^2} \sum_{i,j=1}^{n} \exp\left(-\frac{1}{2}\|(x_i - x_j)\|^2 a^2\right) = T_N, \qquad (2.2.1)$$

say. When $W_a(t) = (2a)^{-d}$ for $t \in [-a,a]^d$, and zero otherwise,

$$T_n = \frac{1}{n^2} \sum_{i \neq j} \prod_{k=1}^{d} \frac{\sin\{a(x_i - x_j)_k\}}{a(x_i - x_j)_k} + \frac{1}{n} = T_U, \qquad (2.2.2)$$

say, where $(x_i - x_j)_k$ stands for the kth component of $(x_i - x_j)$.

Proof. Note that the characteristic function of the normal distribution $N(0, a^2 I_d)$ is $\exp(-a^2\|x\|^2/2)$. Hence

$$\int \left\| \int e^{it'x} dF_{nx}(x) \right\|^2 W_a(t)\, dt$$

$$= \frac{1}{n^2} \sum_{j,k=1}^{n} \int e^{it'(x_j - x_k)} (2\pi a)^{-d/2} \exp\left(-\|t\|^2/2a^2\right) dt$$

$$= \frac{1}{n^2} \sum_{j,k=1}^{n} \exp\left(-\frac{1}{2}\|(x_j - x_k)\|^2 a^2\right).$$

For (2.2.2), recalling that $\cos(x)\cos(y) + \sin(x)\sin(y) = \cos(x - y)$, and noticing that since the uniform distribution on $[-a,a]^d$ is symmetric, its characteristic function is real. Then

$$T_n = \int \left\| \int e^{it'x} dF_{nx}(x) \right\|^2 W_a(t)\, dt$$

$$= \left\{ \int \left| \int \cos(t'x) dF_{nx}(x) \right|^2 W_a(t)\, dt + \int \left| \int \sin(t'x) dF_{nx}(x) \right|^2 W_a(t)\, dt \right\}$$

$$= \frac{1}{n^2} \sum_{j,k=1}^{n} (2a)^{-d} \int_{[-a,a]^d} \cos\{t'(x_j - x_k)\}\, dt$$

$$= \frac{1}{n^2} \sum_{j,k=1}^{n} \int_{[-1,1]^d} \cos\{t'a(x_j - x_k)\}\, dt$$

$$= \frac{1}{n^2} \sum_{j,k=1}^{n} \mathrm{Re}\left[\int_{[-1,1]^d} \exp\{it'a(x_j - x_k)\}\, dt \right]$$

$$= \frac{1}{n^2} \sum_{j,k=1}^{n} \mathrm{Re}\left(\prod_{l=1}^{d} \int_{[-1,1]^d} \exp[it_l\{a(x_j - x_k)\}_l 9\, dt_l \right)$$

$$= \frac{1}{n^2} \sum_{j,k=1}^{n} \mathrm{Re}\left(\prod_{l=1}^{d} \int_{[-1,1]^d} \exp[it_l\{a(x_j - x_k)\}_l]\, dt_l \right)$$

$$= \frac{1}{n^2} \sum_{i \neq j} \prod_{k=1}^{d} \frac{\sin\{a(x_i - x_j)_k\}}{a(x_i - x_j)_k} + \frac{1}{n},$$

where $\mathrm{Re}(\cdot)$ denotes the real part of a complex number.   □

REMARK 2.2.1 *Estimate μ consistently by $\hat{\mu}$, such as the sample mean or sample median. Generate data $y_1', \ldots y_n'$ by the Monte Carlo method and let $x_i' = y_i' \cdot h(x_i - \hat{\mu})$. A test statistic T, say, T_N or T_U, can be computed given $\{z_1, \ldots, z_n\}$. Hence, $T_1, \ldots, T_m$ can be obtained by the Monte Carlo procedure. The test procedure can follow that in Section 2.2 exactly. As $\hat{\mu}$ is a consistent estimator of μ, the test is then asymptotically valid.*

2.3 Simulations and Application

2.3.1 Preamble

In this section, simulations provide evidence about the performance of NMCT in conjunction with the test statistics T_N and T_U for reflection symmetry, elliptical symmetry, Liouville-Dirichlet and symmetric scale mixture. We also make a comparison with bootstrap tests. The application to a real data example is illustrated.

To demonstrate the performance of NMCT, we considered a comparison with the bootstrap test:

$$T^* = \int \left\| \hat{\varphi}_x^*(t) - \hat{\varphi}_x(t) - [\hat{E}_{z^*}^*\{\varphi_Y(t'z^*)\} - \hat{E}_z\{\varphi_Y(t'z)\}] \right\|^2 W_a(t)\, dt,$$

where $\hat{\varphi}_x^*(t)$ is the empirical characteristic function based on the bootstrap data $\{x_1^*, \cdots, x_n^*\}$ from $\{x_1, \cdots, x_n\}$, $\hat{\varphi}_x(t)$ is that based on $\{x_1, \cdots, x_n\}$, $\hat{E}_{z^*}^*\{\varphi_Y(t'z^*)\}$ is the sample mean of $\varphi_Y(t'z_j^*)$ with the bootstrap data $\{z_1^*, \cdots, z_n^*\}$ from $\{z_1, \cdots, z_n\}$, and $\hat{E}_z\{\varphi_Y(t'z)\}$ is that based on $\{z_1, \cdots, z_n\}$. We define T^* as T_N^* and T_U^* respectively, with the normal and uniform weights.

2.3.2 Simulations

Throughout the simulations, the nominal level was $\alpha = 0.05$ and the sample sizes were $n = 10$ and 20. The dimensions of the random vector X were $d = 2, 4$ and 6. For each sample, the p-value was determined using 1000 replications of the Monte Carlo procedures. The empirical powers were estimated as the proportions of times out of 1000 that each procedure rejected the null hypothesis. The random vector had a distribution $N \star \{b(\chi_1^2 - 1)\}$, the convolution of two distributions where N was a hypothetical distribution in accordance with the null hypothesis and $\chi_1^2 - 1$ was d-dimensional distribution whose marginals are the centered chi-squared distribution with one degree of freedom. In the simulations, $b = 0$ and $b = 1$ correspond respectively to the null hypothesis and the alternative. We also investigated the effect of the parameter in the weight function $W_a(\cdot)$ on the test statistics T_N and T_U. We considered $a = 0.5, 2, 3.5, 5$ and 7.5. The unknown location was estimated by the sample mean.

Case (a). Testing for spherical symmetry.

Let $D = N$ be the standard normal $N(0, I_d)$. Clearly the data have zero mean, but we analyzed them as if the location were both known and unknown. Figure 2.1 shows how the power of the tests T_U and T_N varies with a. We only present the results for cases of known location. The results for bootstrap tests and for unknown location are similar. For T_U $a = 2$ seems a good choice, while for T_N $a = 5$ may be better. The power of T_U is quite low while that of T_N may be worthwhile.

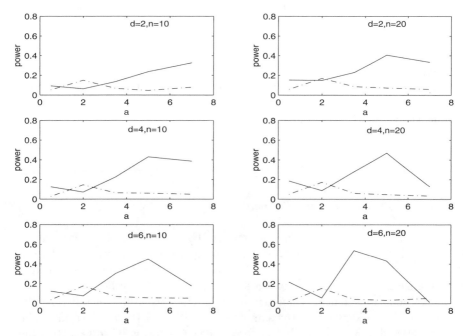

Fig. 2.1. The estimated rejection rate of NMCT for spherical distribution. Solid line is for normal weight and dashed line for uniform weight; a is the parameter in the weight.

In Table 2.1, we report the empirical powers of T_N and T_N^* with $a = 5$.

Table 2.1. *Spherical symmetry: Empirical power with $a = 5$*

		Test	known location		unknown location	
d	n	statistic	N	$N \star (\chi^2 - 1)$	N	$N \star (\chi^2 - 1)$
2	10	T_N	0.048	0.237	0.033	0.237
	10	T_N^*	0.071	0.213	0.028	0.208
	20	T_N	0.053	0.407	0.047	0.410
	20	T_N^*	0.061	0.393	0.039	0.397
4	10	T_N	0.050	0.430	0.037	0.427
	10	T_N^*	0.069	0.421	0.029	0.410
	20	T_N	0.048	0.470	0.043	0.463
	20	T_N^*	0.061	0.447	0.040	0.458
6	10	T_N	0.062	0.450	0.053	0.440
	10	T_N^*	0.074	0.442	0.045	0.430
	20	T_N	0.044	0.430	0.053	0.420
	20	T_N^*	0.063	0.442	0.044	0.426

With a very small sample size $n = 10$, the size of T_N in the known location case is close to the nominal level but the size of T_N with the unknown location is sometimes lower. When $n = 20$, the situation improves. This is the case also for the bootstrap test T_N^*, but T_N^* does worse at achieving the significance level and has lower power than T_N. It is encouraging that both the Monte Carlo test and the bootstrap test are little affected by the dimensionality of the data although twenty data points are clearly scattered sparsely in 6-dimensional space. The dimensionality and the nuisance location parameter do not seem to be very serious problems in this example.

Case (b). Testing for reflection symmetry.

Let $D = N$ be the standard normal $N(0, I_d)$. The data have zero mean, but similar to Case (a), we analyzed them as if the mean were both known and unknown. Figure 2.2 presents the plots of the power functions. Looking at Figure 2.2, the power against a shows that $a = 2$ may be a good choice for the uniform weight function. However for the normal weight function, the choice of a may depend on the sample size. With $n = 10$, $a = 5$ seems to be suitable, and $a = 2$ may be better for $n = 20$. In contrast with Case (a), the power of T_U is good, and even better than that of T_N.

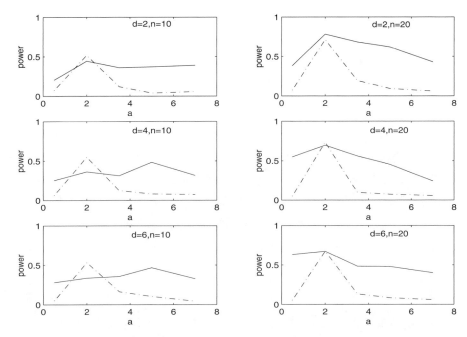

Fig. 2.2. The estimated rejection rate of NMCT for reflectively symmetric distribution. Solid line is for normal weight and dashed line for uniform weight; a is the parameter in the weight.

Tables 2.2 and 2.3 show the empirical powers of T_U and T_N with $a = 2$ and $a = 5$. As in Case (a), the dimensionality may not be a major factor on the performance of the tests. In maintaining the level of significance, both T_N and T_U have similar performance. But the powers of the tests decrease as the dimension of the variable increases. The power of the tests with a known location is slightly higher than that with a nuisance location. The bootstrap tests have similar behavior to that described above, but, overall, perform slightly worse.

Table 2.2 *Reflection symmetry for know location: Empirical power*

		N	$N \star (\chi^2 - 1)$	a = 2	N	$N \star (\chi^2 - 1)$
$d = 2$	$n = 10$, T_N	0.056	0.370	$(a=5)$ $n = 10$, T_U	0.030	0.515
	$n = 10$, T_N^*	0.072	0.365	$(a=5)$ $n = 10$, T_U^*	0.040	0.507
	$n = 20$, T_N	0.056	0.783	$(a=2)$ $n = 20$, T_U	0.047	0.714
	$n = 20$, T_N^*	0.067	0.774	$(a=2)$ $n = 20$, T_U^*	0.042	0.701
$d = 4$	$n = 10$, T_N	0.061	0.483	$(a=5)$ $n = 10$, T_U	0.053	0.550
	$n = 10$, T_N^*	0.068	0.472	$(a=5)$ $n = 10$, T_U^*	0.062	0.527
	$n = 20$, T_N	0.040	0.697	$(a=2)$ $n = 20$, T_U	0.060	0.733
	$n = 20$, T_N^*	0.044	0.702	$(a=2)$ $n = 20$, T_U^*	0.077	0.712
$d = 6$	$n = 10$, T_N	0.030	0.467	$(a=5)$ $n = 10$, T_U	0.037	0.533
	$n = 10$, T_N^*	0.041	0.461	$(a=5)$ $n = 10$, T_U^*	0.044	0.522
	$n = 20$, T_N	0.053	0.470	$(a=2)$ $n = 20$, T_U	0.064	0.673
	$n = 20$, T_N^*	0.059	0.464	$(a=2)$ $n = 20$, T_U^*	0.074	0.659

Table 2.3. *Reflection symmetry with unknown location: Empirical power*

		N	$N \star (\chi^2 - 1)$	a = 2	N	$N \star (\chi^2 - 1)$
$d = 2$	$n = 10$, T_N	0.050	0.296	$(a=5)$ $n = 10$, T_U	0.031	0.383
	$n = 10$, T_N^*	0.058	0.299	$(a=5)$ $n = 10$, T_U^*	0.038	0.380
	$n = 20$, T_N	0.051	0.707	$(a=2)$ $n = 20$, T_U	0.057	0.597
	$n = 20$, T_N^*	0.049	0.701	$(a=2)$ $n = 20$, T_U^*	0.064	0.589
$d = 4$	$n = 10$, T_N	0.043	0.420	$(a=5)$ $n = 10$, T_U	0.043	0.430
	$n = 10$, T_N^*	0.049	0.407	$(a=5)$ $n = 10$, T_U^*	0.045	0.433
	$n = 20$, T_N	0.054	0.590	$(a=2)$ $n = 20$, T_U	0.062	0.617
	$n = 20$, T_N^*	0.056	0.568	$(a=2)$ $n = 20$, T_U^*	0.068	0.607
$d = 6$	$n = 10$, T_N	0.049	0.433	$(a=5)$ $n = 10$, T_U	0.056	0.530
	$n = 10$, T_N^*	0.053	0.430	$(a=5)$ $n = 10$, T_U^*	0.072	0.520
	$n = 20$, T_N	0.059	0.410	$(a=2)$ $n = 20$, T_U	0.065	0.643
	$n = 20$, T_N^*	0.058	0.397	$(a=2)$ $n = 20$, T_U^*	0.077	0.644

Case (c). Testing for Liouville-Dirichlet.

Let the hypothetical distribution $D = L$ be an exponential distribution with the density function $\exp(-\sum_{i=1}^{d} x^{(i)})$ with $x^{(i)} \geq 0$, $i = 1, \ldots, d$ where the $x^{(i)}$ are the components of X. In this case Dirichlet distribution $D(\alpha)$ is with parameter $\alpha = (1, 1, \cdots, 1)$. Figure 2.3 showing the power against a suggests that, in the weight function $W_a(\cdot)$, $a = 0.50$ may be a good choice for both T_N and T_U.

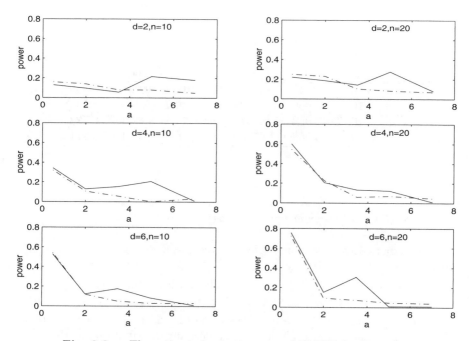

Fig. 2.3. The estimated rejection rate of NMCT for Liouville-Dirichlet distribution. Solid line is for normal weight and dashed line for uniform weight; a is the parameter in the weight.

The power reported in Table 2.4 shows that all the tests have higher power with larger dimension. On the other hand, the tests T_N and T_U maintain the significance level well in most cases but the bootstrap tests T_N^* and T_U^* do not.

Table 2.4. *Liouville distribution: Empirical power with* $a = 0.5$

		L	$L \star (\chi^2 - 1)$		L	$L \star (\chi^2 - 1)$
	$n = 10, T_N$ 0.076		0.133	$n = 10, T_U$ 0.056		0.161
$d = 2$	$n = 10, T_N^*$ 0.081		0.137	$n = 10, T_U^*$ 0.070		0.157
	$n = 20, T_N$ 0.063		0.200	$n = 20, T_U$ 0.053		0.250
	$n = 20, T_N^*$ 0.063		0.201	$n = 20, T_U^*$ 0.067		0.233
	$n = 10, T_N$ 0.043		0.343	$n = 10, T_U$ 0.060		0.322
$d = 4$	$n = 10, T_N^*$ 0.061		0.322	$n = 10, T_U^*$ 0.067		0.324
	$n = 20, T_N$ 0.060		0.601	$n = 20, T_U$ 0.056		0.543
	$n = 20, T_N^*$ 0.065		0.580	$n = 20, T_U^*$ 0.064		0.547
	$n = 10, T_N$ 0.046		0.526	$n = 10, T_U$ 0.051		0.540
$d = 6$	$n = 10, T_N^*$ 0.060		0.508	$n = 10, T_U^*$ 0.058		0.531
	$n = 20, T_N$ 0.063		0.757	$n = 20, T_U$ 0.043		0.723
	$n = 20, T_N^*$ 0.068		0.740	$n = 20, T_U^*$ 0.054		0.711

Case (d). Testing for symmetric scale mixture.

The hypothetical distribution $D = S$ corresponds to the density function $\frac{1}{2^d} \exp(-\sum_{i=1}^{d} |x^{(i)}|)$. Let $g(X) = \sum_{i=1}^{d} |X^{(i)}|$; the density of X is a product of $\exp(-\sum_{i=1}^{d} |x^{(i)}|)$ and a constant on the domain $C^d = \{u : u \in R^d, \sum_{i=1}^{d} |u^{(i)}| = 1\}$. Hence, $\sum_{i=1}^{d} |X^{(i)}|$ is independent of $X / \sum_{i=1}^{d} |X^{(i)}|$. The mean is zero. We also regard it as a known and an unknown location. NMCT was applied to test whether the underlying distribution has the form $f(\sum_{i=1}^{d} |x^{(i)}|)$. Figure 2.4 reporting the plot of power versus a suggests that, for T_U, a good choice of a is 2, and for T_N, with normal weight, $a = 3.5$ may be good. The power of T_U is quite low so that the normal weight seems preferable here.

In this case the dimensionality of the variable influences the power performance. The significance level is maintained well by T_N, but the bootstrap test T_N^* has larger size than the nominal level in most cases. However, the tests seem not to be sensitive to the alternative. The choice of the weight function deserves further study.

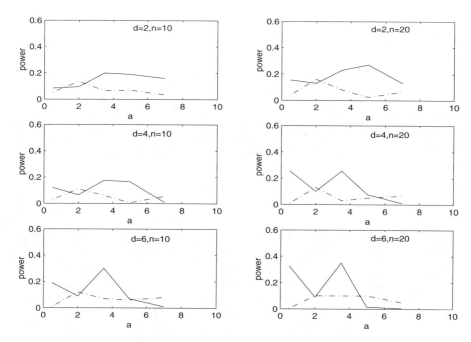

Fig. 2.4.   The estimated rejection rate of NMCT for symmetric scale mixture distribution. Solid line is for normal weight and dashed line for uniform weight; a is the parameter in the weight.

Table 2.5. *Symmetric scale mixture: Empirical power with a = 3.5*

		known location		unknown location	
		S	$S \star (\chi^2 - 1)$	S	$S \star (\chi^2 - 1)$
$d = 2$	$n = 10,\ T_N$	0.055	0.160	0.060	0.150
	$n = 1,\ T_N^*$	0.058	0.162	0.062	0.147
	$n = 20,\ T_N$	0.050	0.231	0.045	0.223
	$n = 20,\ T_N^*$	0.054	0.234	0.048	0.227
$d = 4$	$n = 10,\ T_N$	0.050	0.232	0.055	0.261
	$n = 10,\ T_N^*$	0.055	0.233	0.058	0.250
	$n = 20,\ T_N$	0.055	0.255	0.057	0.313
	$n = 20,\ T_N^*$	0.056	0.251	0.060	0.301
$d = 6$	$n = 10,\ T_N$	0.061	0.235	0.062	0.292
	$n = 10,\ T_N^*$	0.059	0.237	0.067	0.279
	$n = 20,\ T_N$	0.045	0.353	0.053	0.369
	$n = 20,\ T_N^*$	0.047	0.360	0.059	0.354

In summary, generally NMCT maintains the level of significance well and outperforms the bootstrap tests. The nuisance location parameter does not have much impact on the performance of NMCT, but the dimensionality of the variable and the choice of weight function may. For instance, the uniform weight function is good in testing reflection symmetry, while the normal weight function is better in testing elliptical symmetry and symmetric scale mixture.

2.3.3 Application

We revisited the dataset presented and examined by Royston (1983) for multi-normality. The six measurements are hemoglobin concentrations, x_1, packed cell volume, x_2, white blood cell count, x_3, lymphocyte count, x_4, neutrophil count, x_5, serum lead concentration, x_6. There are 103 observations. Before the analysis, x_3, x_4, x_5 and x_6 were logarithmically transformed. Royston employed the Shapiro-Wilks W test for normality of the data and concluded that all the six marginal distributions may be univariate normal. However, there is evidence (Royston, 1983) to suggest that the transformed squared radii are jointly non-normal. We used T_N with $a = 2$ to check whether or not the measurements are jointly reflectively symmetric and jointly spherically symmetric. One thousand Monte Carlo simulations were carried out for calculating p-values. The estimated p-values are 0.149 and 0.141, which do not provide convincing evidence against the null hypotheses of these two types of symmetry. On the other hand, the scatter plot of x_3 against x_4 shows a remarkable linear relationship and the correlation coefficient of these two variables is 0.61. Based on the measurements except x_3, the p-values of the test

T_N for reflection and elliptical symmetries are respectively 0.67 and 0.71. The symmetries are tenable.

3

Asymptotics of Goodness-of-fit Tests for Symmetry

3.1 Introduction

In chapter 2, we proposed the goodness-of-fit tests for four kinds of multivariate distributions. Because of the importance of symmetric distributions in practice, testing for elliptically symmetric distribution and reflectively symmetric distribution have received more attentions than the other two types of distributions. Therefore, we consider more studies of tests for these two types of distributions. Especially, we study the asymptotic behavior of tests. In this chapter, we construct tests that are based on characteristic functions of distribution. The elliptical symmetry is a generalization of the spherical symmetry that was discussed in Chapter 2. Most of materials in this chapter are from Zhu and Neuhaus (2003).

DEFINITION 3.1.1 *Let X be a d-dimensional random vector. The distribution of X is said to be elliptically symmetric with a center $\mu \in R^d$ and a matrix A if for all orthogonal d by d matrices H the distributions of $HA(X - \mu)$ are identical. A is called the shape matrix.*

It is easy to prove that when A is the identity matrix and $\mu = 0$, the distribution of X is spherically symmetric. Readers can refer to Fang, Kotz and Ng (1990). Throughout this chapter, we assume that the covariance matrix Σ of X is positive definite. In this case, A is equal to $\Sigma^{-1/2}$ and $\Sigma^{-1/2}(X - \mu)$ is spherically symmetrically distributed if X has an elliptically symmetric distribution.

Symmetric distributions play crucial roles in multivariate data analysis. Elliptically symmetric distribution (elliptical distribution for short) possesses many nice properties which are analogous to those of multivariate normal distribution. Hence, if one knows that the variable is elliptically symmetrically distributed, some tools for classical multivariate analysis, as Friedman (1987) pointed out, may still be applicable for analyzing the data. Additionally, the dimensional reduction techniques have been developed in recent years for overcoming the problem of dimensionality in data analysis, one of which is Sliced

Inverse Regression (SIR) (see Li, 1991). The most important subclass of distributions satisfying the designed condition of SIR is just the one of elliptical distributions. Consequently, testing elliptical symmetry is important and relevant in multivariate analysis. The hypothesis to be tested is that whether the underlying distribution of X is elliptically symmetric or not. Further, if we have reflection symmetry of a distribution, which is defined in Chapter 2, the SIR can also be approximately used.

3.2 Test Statistics and Asymptotics

3.2.1 Testing for Elliptical Symmetry

The null hypothesis is, for all orthogonal matrices H,

$$H_0 : \ \Sigma^{-1/2}(X - \mu) = H\Sigma^{-1/2}(X - \mu) \quad \text{in distribution.}$$

That is, we want to test whether $\Sigma^{-1/2}(X - \mu)$ is spherically symmetrically distributed. The spherical symmetry implies that the imaginary part of the characteristic function of $\Sigma^{-1/2}(X - \mu)$ is zero, that is,

$$E\big(\sin(ta^\tau \Sigma^{-1/2}(X - \mu))\big) = 0$$

for all $t \in R^1$ and $a \in S^d = \{a : ||a|| = 1, a \in R^d\}$. The empirical version of this imaginary part is $P_n\{\sin(ta^\tau \hat{A}(X - \hat{\mu}))\}$ where P_n is the empirical probability measure based on the sample points $\{X_1, \cdots, X_n\}$ which are $i.i.d.$ copies of X, $P_n(f)$ stands for $(1/n)\sum_{j=1}^n f(X_j)$ for each function $f(\cdot)$, $\hat{A} = \Sigma^{-1/2}$ or $= \hat{\Sigma}^{-1/2}$, the sample covariance and $\hat{\mu} = \mu$ or $= \bar{X}$ the sample mean respectively in accordance with the parameters being known or unknown.

A test statistic is defined as

$$\int_{S^d} \int_I (\sqrt{n} P_n\{\sin(ta^\tau \hat{A}(X - \hat{\mu}))\})^2 w(t) dt d\nu(a), \tag{3.2.1}$$

where $w(\cdot)$ is a weight function with a compact support, ν is the uniform distribution on S^d and I is a working region. In this chapter, we consider that I is a compact subset of the real line R. The null hypothesis H_0 is rejected for the large values of the test statistic.

In order to study the asymptotic properties of the test statistic, we define an empirical process by

$$V_n = \{V_n(\mathbf{X}_n, \hat{\mu}, \hat{A}, t, a) = \sqrt{n} P_n\{\sin(ta^\tau \hat{A}(X - \hat{\mu}))\} : (t, a) \in I \times S^d\}, \tag{3.2.2}$$

and the test statistic in (3.2.1) can be rewritten as

$$T_n = \int_{S^d} \int_I \{V_n(\mathbf{X}_n, \hat{\mu}, \hat{A}, t, a)\}^2 dw(t) d\nu(a). \tag{3.2.3}$$

The asymptotic behavior, under the null hypothesis, of the empirical process defined above is presented in the following theorem and corollary. For simplicity, assume a Gaussian process with the index set $I \times S^d$ is continuous if its sample paths are bounded and uniformly continuous with respect to $(t, a) \in I \times S^d$.

THEOREM 3.2.1 *Assume that* $P\{X = \mu\} = 0$ *and* $E||X - \mu||^4 < \infty$. *Then under* H_0

1). If the center μ *is given and then* $\hat{\mu} = \mu$, *the process* V_n *converges in distribution to a centered continuous Gaussian process* $V_1 = \{V_1(t, a) : (t, a) \in I \times S^d\}$ *with the covariance kernel: for* $(t, a), (s, b) \in I \times S^d$

$$E\{\sin(ta^\tau A(X - \mu)) \sin(sb^\tau A(X - \mu))\}.$$

(3.2.4)

2). If the center μ *is an unknown parameter and then* $\hat{\mu} = \bar{X}$. *Let*

$$k(t, a, x) = \sin(ta^\tau A(x - \mu)) - ta^\tau A(x - \mu) E(\cos(ta^\tau A(X - \mu))). \quad (3.2.5)$$

The process V_n *converges in distribution to a centered continuous Gaussian process* $V_2 = \{V_2(t, a) : (t, a) \in I \times S^d\}$ *with the covariance kernel: for* $(t, a), (s, b) \in I \times S^d$,

$$E\{k(t, a, x)k(s, b, x)\}.$$

(3.2.6)

The convergence of the test statistic is a direct consequence of Theorem 2.1.

COROLLARY 3.2.1 *The test statistics* T_n *that are associated with known and unknown centers converge in distribution to the quadratic functionals*

$$\int_{S^d} \int_I V_1^2(t, a) d\,w(t)\, d\nu(a) \quad and \quad \int_{S^d} \int_I V_2^2(t, a) d\,w(t)\, d\nu(a)$$

respectively.

We now investigate the behavior of the test under alternatives. For convenience, let $\sin^{(i)}(c)$ be i-th derivative of $\sin(\cdot)$ at point c. If there is a direction $a \in S^d$ such that $E[\sin(ta^\tau A(X - \mu))] \neq 0$ for some $t \in I$, it is easily derived that from the continuity of function $E[\sin(ta^\tau A(X - \mu))]$ w.r.t. (t, a), the test statistic T_n converges in distribution to infinity since the process V_n converges in distribution to infinity. This means that the tests are consistent against global alternatives. The rest of this section focuses on the investigation with local alternatives.

Suppose that the *i.i.d.* d-dimensional vectors $X_i = X_{in}$ have the expression $Z_i + Y_i / n^\alpha, i = 1, \cdots, n$ for some $\alpha > 0$. The center $\mu = \mu_n = E(Z) + E(Y)/n^\alpha$. When Z_i is independent of Y_i, the distribution of X_{in} is a convolution of two distributions, and one of them converges to the degenerate distribution at zero with the rate n^α in certain sense.

THEOREM 3.2.2 *Assume that the following conditions hold:*

1) *Both distributions of Z and of Y are continuous. In addition, Z is elliptically symmetric with the center $E(Z)$ and the shape matrix Σ.*
2) *There is an integer l being the smallest one such that*

$$\sup_{(t,a)\in I\times S^d} |B_l(t,a)|$$

$$=: \sup_{(t,a)\in I\times S^d} |E((ta^\tau A(Y - E(Y)))^l \sin^{(l)}(ta^\tau A(Z - E(Z))))| \neq 0,$$

$$E(||Y||^{2l}) < \infty, \qquad and \qquad E(||Y||^{2(l-1)}||Z||^2) < \infty. \qquad (3.2.7)$$

Then when $\alpha = 1/(2l)$, if $\hat{\mu} = \mu$

$$T_n \Longrightarrow \int_{S^d}\int_I (V_1(t,a) + B_l(t,a)/l!)^2 d\,w(t)d\,\nu(a), \qquad (3.2.8)$$

and if $\hat{\mu} = \bar{X}$,

$$T_n \Longrightarrow \int_{S^d}\int_I (V_2(t,a) + B_l(t,a)/l!)^2 d\,w(t)d\,\nu(a). \qquad (3.2.9)$$

where " $\Longrightarrow$ " stands for the convergence in distribution, V_1 and V_2 are the Gaussian processes defined in Theorem 3.2.1.

REMARK 3.2.1 *Comparing the limits under the null hypothesis in Corollary 3.2.1 with those under the alternative in Theorem 3.2.2, we see that the test can detect the local alternatives distinct $O(n^{-1/(2l)})$ from the null. In certain cases, this rate can achieve $O(n^{-1/2})$, that is, $l = 1$. For example, suppose that Z has the uniform distribution on S^d and $Y = (Z_1^2 - 1, \cdots, Z_d^2 - 1)$, we can see easily that, via a little elementary calculation, $\sup_{(t,a)\in I\times S^d} |E(ta^\tau AY \cos(ta^\tau AZ))| \neq 0$. In contrast, when Z and Y are independent, $l \geq 3$, namely, the test can detect, at most, alternatives distinct $O(n^{-1/6})$ from the null. In fact, it is clear that for $l = 1, 2$*

$$\sup_{(t,a)\in I\times S^d} |E((ta^\tau AY)^l \sin^{(l)}(ta^\tau AZ))| = 0.$$

3.2.2 Testing for Reflection Symmetry

As defined in Chapter 2, a d-variate random variable X is said to be reflectively symmetric about a center μ if

$$(X - \mu) \text{ and } -(X - \mu) \text{ have the same distribution.}$$

Note that it is equivalent to the imaginary part of the characteristic function of $X - \mu$ being equal zero, i.e.,

$$E\{\sin(t^\tau(X - \mu))\} = 0 \qquad for \qquad t \in R^d. \qquad (3.2.10)$$

Let $X_1, \ldots, X_n$ be *i.i.d.* copies of X and $P_n(\cdot)$ the corresponding empirical probability measure. Based on (3.2.10), the test statistic can have a similar form to that for elliptical symmetry:

$$Q_2 = n \int_A \{P_n(\sin(t^\tau(X - \hat{\mu})))\}^2 dw(t)$$

where A (a general region) are working regions, $w(\cdot)$ is a distribution function on R^d, and $P_n(f(x))$ is defined in Subsection 3.2.1. We now use other notations to represent Q_2. Define an empirical process

$$\{U_{n1}(\mathbf{X}_n, \hat{\mu}, t) = \sqrt{n} P_n\{\sin(t^\tau(X - \hat{\mu}))\} : t \in A\}, \qquad (3.2.11)$$

where $\mathbf{X}_n = (X_1, \ldots, X_n)$, $\hat{\mu}$ is an estimate of μ when μ is unknown. The test statistic is then rewritten as

$$Q_2(\mathbf{X}_n, \hat{\mu}) = \int_A [U_{n1}(\mathbf{X}_n, \hat{\mu}, t)]^2 dw(t), \qquad (3.2.12)$$

Heathcote, Rachev and Cheng (1995, Theorem 3.2) have obtained the convergence of U_{n1} to a Gaussian process U under the null hypothesis that in distribution, $X - \mu = -(x - \mu)$. The convergence of $Q_2(\mathbf{X}_n, \hat{\mu})$ is a direct consequence of their result. Therefore, we only investigate the behavior of the test and the NMCT's under s.

For convenience, let $\sin^{(i)}(t^\tau X)$ be i-th derivative of $\sin(\cdot)$ at the point $t^\tau X$. Suppose that *i.i.d.* d-dimensional variables have the representation $Z_i + y_i/n^\alpha, i = 1, \ldots, n$, for some $\alpha > 0$. This means that the distribution of x is the convolution of a symmetric distribution and a distribution converging to the degenerate one. The following theorem reveals the power behavior of the tests for such local alternatives.

THEOREM 3.2.3 *Assume that the following conditions hold:*

1) Both distributions of Z and of y are continuous and, in addition, Z is reflectively symmetric about a known center μ.

2) Let l denote the smallest integer, such that

$$\sup_{t \in A} |B_l(t)| := \sup_{t \in A} |E(t^\tau(y - Ey)^l \sin^{(l)}(t^\tau(Z - EZ)))| \neq 0,$$

$$E(||y||^{2l}) < \infty, \qquad and \qquad E(||y||^{2(l-1)}||Z||^2) < \infty. \qquad (3.2.13)$$

Then

$$\{\sqrt{n} P_n\{\sin(t^\tau(Z + y/n^{1/(2l)} - EZ - Ey/n^{1/(2l)}))\} : t \in A\}$$
$$= \{\sqrt{n} P_n\{\sin(t^\tau(Z - EZ)) + (1/l!)B_l(t)\} : t \in A\} + o_p(1). \qquad (3.2.14)$$

This leads to convergence in distribution ($\Longrightarrow$)

$$\int_A \left\{\sqrt{n} P_n(\sin(t^\tau(Z + y/n^{1/(2l)}) - EZ - Ey/n^{1/(2l)})))\right\}^2 dw(t)$$

$$\implies \int_A (U(t) + (1/l!)B_l(t))^2 dw(t), \tag{3.2.15}$$

where $\{U(t) : t \in A\}$ is a Gaussian process defined in Heathcote, Rachev and Cheng (1995, Theorem 3.2).

REMARK 3.2.2 It is worthwhile to note that the results of this theorem are similar to that of Theorem 3.2.2. We can also conclude that the test can detect local alternatives converging to the null hypothesis at $n^{1/(2l)}$-rate or slower (that is, the test statistic will converge in distribution to infinity under the local alternative distinct with slower convergence rate from the null). This rate can also achieve $O(n^{-1/2})$, that is, $l = 1$. For example, if Z has a uniform distribution on $[-\sqrt{3}, \sqrt{3}]^d$ and if $y = (z_1^2 - 1, \cdots, z_d^2 - 1)$, we can also see that $\sup_{t \in [-1,1]^d} |E(t^\tau y \cos(t^\tau Z))| \neq 0$. Hence, $l = 1$. Similar to the testing for elliptical symmetry, when Z and y are independent of each other, l is at least three, and the tests can detect, at most, alternatives converging to the null hypothesis at $n^{1/6}$-rate. In fact, it is clear that for $l = 1, 2$

$$\sup_{t \in [-1,1]^d} |E((t^\tau y)^l \sin^{(l)}(t^\tau Z))| = 0.$$

3.3 NMCT Procedures

3.3.1 NMCT for Elliptical Symmetry

Since under the null hypothesis $A(X - \mu)$ is spherically symmetrically distributed, we have

$$A(X - \mu) = U \bullet \|A(X - \mu)\| \quad \text{in distribution} \tag{3.3.1}$$

where $U = A(X - \mu)/\|A(X - \mu)\|$ is uniformly distributed on S^d and independent of $\|A(X - \mu)\|$ (see, e.g. Dempster (1969)). Hence for any U being uniformly distributed on S^d, $U \bullet \|A(X - \mu)\|$ has the same distribution as $A(X - \mu)$. When μ and A are known, the situation is the same as that discussed in Chapter 2: we can have the exact validity of the test.

It is worthwhile to point out that when both μ and Σ are unknown, the NMCT procedure needs some modification when the test statistic T_n is used. For the convenience of comparison, we still present the algorithm with known μ and A.

Step 1. Generate by computer i.i.d. random vectors, say u_i, of size n with uniform distribution on S^d, let $\mathbf{U}_n = (u_1, \cdots, u_n)$. The new data are $u_i \bullet \|A(X_i - \mu)\|$.

Step 2. Accordingly as the empirical process defined in (3.2.2), we define a conditional empirical process. For fixed $\mathbf{X}_n = (X_1, \cdots, X_n)$, let

$$V_{n1}(\mathbf{U}_n)$$
$$= \{V_{n1}(\mathbf{X}_n, \mathbf{U}_n, t, a) = \sqrt{n} P_n \{\sin(ta^\tau u \| A(X - \mu) \|)\} : (t, a) \in I \times S^d\},$$
$$(3.3.2)$$

and calculate the value of the statistic

$$T_{n1}(\mathbf{U}_n) = \int_{S^d} \int_I \{V_{n1}(\mathbf{X}_n, \mathbf{U}_n, t, a)\}^2 dw(t) d\nu(a). \qquad (3.3.3)$$

Step 3. Repeat steps 1 and 2 m times to obtain m values $E_{n1}(\mathbf{U}_n^{(j)}), j = 1, \cdots, m$.

Step 4. Define $E_{n1}(U_n^{(0)})$ as the value of E_n. Estimate the p-value by $p = k/(m + 1)$ where k is the number that $E_{n1}(U_n^{(j)})$ $j = 0, 1, \cdots, m$ are greater than or equal to $E_{n1}(U_n^{(0)})$.

From Proposition 1.2.1, we can obtain the exact validity of the test.

When the center μ is known but the shape matrix A is unknown, we still use the above algorithm except replacing A by its estimator $\hat{A} = \hat{\Sigma}^{-1/2}$. When μ is unknown, the situation is not so simple. This is different from that of Chapter 2 because of the use of a different test statistic. In order to ensure the equivalence between the conditional empirical process below and its unconditional counterpart, we shall use the following fact to construct conditional empirical process. It can be derived by the triangle identity and $P_n X = \bar{X}$ that uniformly on $t \in I$ and $a \in S^d$

$$\sqrt{n} P_n (\sin(ta^\tau \hat{A}(X - \bar{X})))$$
$$= \sqrt{n} \Big(P_n(\sin(ta^\tau \hat{A}(X - \mu))) \Big) \Big(\cos(ta^\tau \hat{A} P_n(X - \mu)) \Big)$$
$$- \sqrt{n} \Big(P_n(\cos(ta^\tau \hat{A}(X - \mu))) \Big) \Big(\sin(ta^\tau \hat{A} P_n(X - \mu)) \Big)$$
$$= \sqrt{n} P_n (\sin(ta^\tau A(X - \mu)))$$
$$- \sqrt{n} \Big(P_n(\cos(ta^\tau A(X - \mu))) \Big) \Big(\sin(ta^\tau A P_n(X - \mu)) \Big) + o_p(1).$$

We then define a conditional empirical process in Step 2 of the algorithm as

$$V_{n2}(\mathbf{U}_n) = \{V_{n2}(\mathbf{X}_n, \mathbf{U}_n, \hat{\mu}, \hat{A}, t, a) : (t, a) \in I \times S^d\}, \qquad (3.3.4)$$

where

$$V_{n2}(\mathbf{X}_n, \mathbf{U}_n, \hat{\mu}, \hat{A}, t, a)$$
$$= \sqrt{n} P_n \{\sin(ta^\tau u \bullet \| \hat{A}(X - \hat{\mu}) \|)\}$$
$$- \sqrt{n} P_n \{\cos(ta^\tau u \bullet \| \hat{A}(X - \hat{\mu}) \|) \sin(ta^\tau P_n(u \bullet \| \hat{A}(X - \hat{\mu}) \|)).$$
$$(3.3.5)$$

The associated conditional statistic is defined as

$$T_{n2}(\mathbf{U}_n) = \int_{S^d} \int_I \{V_{n2}(\mathbf{X}_n, \mathbf{U}_n, \hat{\mu}, \hat{A}, t, a)\}^2 dw(t) d\nu(a). \qquad (3.3.6)$$

In the following theorem, we present the asymptotic equivalence between the conditional empirical processes $V_{n1}(\mathbf{U}_n)$ and $V_{n2}(\mathbf{U}_n)$ and their unconditional counterparts. The asymptotic validity of $T_{n1}(\mathbf{U}_n)$ and $T_{n2}(\mathbf{U}_n)$ is a direct consequence.

THEOREM 3.3.1 *Assume, in addition to the conditions of Theorem 3.2.1, that $P\{X = \mu\} = 0$. Then the conditional empirical processes $V_{n1}(\mathbf{U}_n)$ and $V_{n2}(\mathbf{U}_n)$ given $\mathbf{X}_n$ in (3.3.2) and (3.3.5) converge, for almost all sequences $\{X_1, \cdots, X_n, \cdots\}$, in distribution to the Gaussian process V_1 and V_2 defined in Theorem 2.1 respectively, which are the limits of the unconditional counterparts V_n with known and unknown centers. This leads up to the conclusion that the conditional statistics $T_{n1}(\mathbf{U}_n)$ and $T_{n2}(\mathbf{U}_n)$ given $\mathbf{X}_n$ in (3.3.3) and (3.3.6) have almost surely the same limits as those of the statistics T_n with known and unknown centers respectively, $T = \int (V(a,t))^2 dw(t) d\nu(a)$ and $T_1 = \int (V_1(a,t))^2 dw(t) d\nu(a)$.*

REMARK 3.3.1 *The optimal choice of the working region I and the weight function $w(\cdot)$ is an interesting problem. But it is worth mentioning that, in some cases, the choice of working regions is not very important. We now show an example in which the fact the imaginary part of the characteristic function being equal zero in a compact subset of R^d such as $[-2, 2] \times S^d$ is equivalent to that the imaginary part is zero in whole space R^d. Suppose that the moment generating function of a multivariate vector, X say, exists in a sphere $[-b, b] \times S^d$, $b > 0$. Then the moment generating function of $a^\tau X$, the linear projector of X on R^1, exists in an interval $[-b_1, b_1]$ for any $a \in S^d$, where b_1 does not depend on a. If the imaginary part of the characteristic function of X equals zero in a sphere $[-b_2, b_2] \times S^d$, so does the imaginary part of the characteristic function of $a^\tau X$ in an interval $[-b_3, b_3]$. It is easy to see that all moments of $a^\tau X$ with odd orders equal zero. This means that the characteristic function of $a^\tau X$ is real, and then $a^\tau X$ is symmetric about the origin for any a. This conclusion implies, in turn, that the imaginary part of the characteristic function of X is zero in R^d. Consequently, the choice of working region is not very important in such a case.*

REMARK 3.3.2 *Romano (1989) proposed a general method of the randomization tests. From the idea of permutation test proposed by Hoeffding (1952), the randomization tests are constructed in terms of the invariance of the distribution for a class $\mathbf{G_n}$ of transformations, see page 151 of Romano (1989). The spherically symmetric distribution has such an invariance property. For testing spherical symmetry, our test procedure is similar to Romano's.*

3.3.2 NMCT for Reflection Symmetry

We use a generic notation where P_n stands for a probability measure which may rest upon different sets of variables for each appearance. Here P_n is the empirical measure of $(e_i, x_i), i = 1, \ldots, n$, where $e_1, \ldots, e_n$ are $i.i.d.$ univariate variables, $e_i = \pm 1, i = 1, \ldots, n$, with probability values one half; define $E_n = (e_1, \ldots, e_n)$. For the known center case define an NMCT process, given $\mathbf{X}_n$, by

$$\{V_n(E_n, \mathbf{X}_n, t) = \sqrt{n} P_n\{\sin(t^\tau e \bullet (X - \mu))\} : t \in A\}. \qquad (3.3.7)$$

The resulting NMCT statistics given $\mathbf{X}_n$ are

$$Q_2(E_n, \mathbf{X}_n) = \int_A (V_n(E_n, \mathbf{X}_n, t))^2 dw(t). \qquad (3.3.8)$$

When the working region A is a cube $[-a, a]^d$ and the weight function $w(\cdot)$ is the uniform distribution on this cube, $Q_2(E_n, \mathbf{X}_n)$ has a specific form which is easy to compute. In fact,

$$\begin{aligned}
Q_2(E_n, \mathbf{X}_n) &= \int_{[-a,a]^d} (V_n(E_n, \mathbf{X}_n, t))^2 dw(t) \\
&= \frac{1}{n} \sum_{i=1}^n \sum_{j=1}^n e_i e_j I(i,j),
\end{aligned} \qquad (3.3.9)$$

where

$$I(i,j) = \frac{1}{2}\Big(\prod_{k=1}^d \frac{\sin(a(X_i - X_j)_k)}{a(X_i - X_j)_k} - \prod_{k=1}^d \frac{\sin(a(X_i + X_j - 2\mu)_k)}{a(X_i + X_j - 2\mu)_k} \Big),$$

and $(x)_k$ means the k-th component of x.

This formula can be justified as follows. Note that

$$\sin(x) \cdot \sin(y) = \frac{1}{2}(\cos(x - y) - \cos(x + y)).$$

Then

$$\begin{aligned}
Q_2(E_n, X_n) &= (2a)^{-d} \int_{[-a,a]^d} \Big\{ \frac{1}{\sqrt{n}} \sum_{i=1}^n \sin(t^\tau e_i \bullet (X_i - \mu)) \Big\}^2 dt \\
&= \frac{1}{n} \sum_{i,j=1}^n \Big\{ (2a)^{-d} \int_{[-a,a]^d} \sin(t^\tau (X_i - \mu)) \sin(t^\tau (X_i - \mu)) dt \Big\} e_i e_j \\
&= \frac{1}{n} \sum_{i=1}^n \sum_{j=1}^n \Big\{ (2)^{-d-1} \int_{[-1,1]^d} \cos(t^\tau a \bullet (X_i - X_j)) \\
&\qquad\qquad - \cos(t^\tau a \bullet (X_i + X_j - 2\mu)) dt \Big\} e_i e_j \\
&:= \frac{1}{n} \sum_{i=1}^n \sum_{j=1}^n e_i e_j I(i,j).
\end{aligned}$$

Denote "Re" the real part of the characteristic function of a distribution. Since the uniform distribution on $[-1, 1]^d$ is symmetric, we have for u having uniform distribution on $[-1, 1]^d$

$$I(i, j) = 2^{-d-1} E(\cos(u^\tau a \bullet (X_i - X_j)) - \cos(u^\tau a \bullet (X_i + X_j - 2\mu)))$$

$$= 2^{-d-1} \left(Re \ E(e^{(u^\tau a \bullet (X_i - X_j))}) - Re \ E(e^{(u^\tau a \bullet (X_i + X_j - 2\mu))}) \right)$$

$$= 2^{-d-1} \left(Re \ \prod_{k=1}^{d} E(e^{(u_k a(X_i - X_j)_k)}) - Re \ \prod_{k=1}^{d} E(e^{(u_k a(X_i + X_j - 2\mu)_k)}) \right)$$

$$= \frac{1}{2} \left(\prod_{k=1}^{d} \frac{\sin(a(X_i - X_j)_k)}{a(X_i - X_j)} - \prod_{k=1}^{d} \frac{\sin(a(X_i + X_j - 2\mu)_k)}{a(X_i + X_j - 2\mu)} \right).$$

The justification is completed.   □

When μ is known, the exact validity of the test of (3.3.8) can follow the generic theory in Chapter 1; we omit the details here.

For the unknown center case, we need a modified version. In order to ensure the equivalence between the conditional empirical process, which will be defined below, and its unconditional counterpart in (3.2.11), both versions cannot be the same. The definition of our conditional empirical process is motivated by the following fact which is similar to that of (3.3.4). For an unknown center μ, an estimate $\hat{\mu}$ is needed. We use $\hat{\mu} = \bar{X}$, the sample mean. It can be proved that uniformly on $t \in A$

$$\sqrt{n} P_n(\sin(t^\tau (X - \bar{X})))$$
$$= \sqrt{n} \Big(P_n(\sin(t^\tau (X - \mu))) \Big) \Big(\cos(t^\tau P_n(X - \mu)) \Big)$$
$$- \sqrt{n} \Big(P_n(\cos(t^\tau (X - \mu))) \Big) \Big(\sin(t^\tau P_n(X - \mu)) \Big)$$
$$= \sqrt{n} P_n(\sin(t^\tau (X - \mu))$$
$$- \sqrt{n} \Big(P_n(\cos(t^\tau (X - \mu))) \Big) \Big(\sin(t^\tau P_n(X - \mu)) \Big) + o_p(1).$$

Accordingly, we define an estimated conditional process $\{V_{n1}(E_n, \mathbf{X}_n, \bar{X}, t) : t \in A\}$ given $\mathbf{X}_n$ by

$$V_{n1}(E_n, \mathbf{X}_n, \bar{X}, t)$$
$$= \sqrt{n} P_n(\sin(t^\tau e \bullet (X - \bar{X})))$$
$$- \sqrt{n} \sin(t^\tau P_n(e \bullet (X - \bar{X}))) P_n(\cos(t^\tau e \bullet (X - \bar{X}))). \quad (3.3.10)$$

The NMCT is defined by

$$Q_2(E_n, \mathbf{X}_n, \bar{X}) = \int_A (V_{n1}(E_n, \mathbf{X}_n, \bar{X}, t))^2 dw(t). \quad (3.3.11)$$

The following theorem states the asymptotic validity of the NMCT Q_2.

THEOREM 3.3.2 *Assume that $X_1, \ldots, X_n, \ldots$ are i.i.d. random variables which are reflectively symmetric about an unknown center μ. Let $E_n^{(1)}, \ldots, E_n^{(m)}, \ldots$ be independent copies of E_n. Then for any $0 < \alpha < 1$,*

$$\lim_{n \to \infty} P\{Q_2(\mathbf{X}_n, \bar{X}) > m - [m\alpha] \text{ of } Q_2(E_n^{(j)}, \mathbf{X}_n, \bar{X})^\tau s\}$$

$$= \lim_{n \to \infty} P\Big\{Q_2(E_n^0, \mathbf{X}_n, \mu) + O_p(1/\sqrt{n}) >$$

$$m - [m\alpha] \text{ of } (Q_2(E_n^{(j)}, \mathbf{X}_n, \mu) + O_p(1/\sqrt{n}))^\tau s\Big\}$$

$$\leq \frac{[m\alpha] + 1}{m + 1}. \qquad (3.3.12)$$

3.3.3 A Simulation Study

When shape matrix is known, Chapter 2 has contained a simulation. In this section, we only consider a simulation with unknown shape matrices. That is, 1) μ is given and Σ is unknown parameter; and 2) both μ and Σ are unknown. The test statistics $T_{ni}, i = 1, 2$ are in accordance with these two cases respectively. The simulation is only for elliptical symmetry. In the simulation results reported below, the sample size $n = 20, 50$. The dimension d of random vector X is $2, 4, 6$. For a power study, we consider the vector $X = Z + b \cdot Y$ with $b = 0.00, 0.25, 0.5, 0.75, 1.00$, and 1.25, where Z has a normal distribution $N(\mu, \Sigma)$ and Y is the random vector with the independent χ_1^2 components. The hypothetical distribution was normal $N(\mu, \Sigma)$. That is, $b = 0.00$ corresponds to the null hypothesis H_0. When $b \neq 0.00$, the distribution is no longer elliptically symmetric. In the simulation, we generated data from $N(0, I_3)$ and, accordingly as different setup, regarded the symmetric center and the shape matrix as known or unknown parameters separately.

To determine critical values when given the data $\{(Y_1, Z_1), \cdots, (Y_n, Z_n)\}$, we generated 1000 $\mathbf{U}_n$ pseudo-random vectors of $n = 20$ and $n = 50$ by the Monte Carlo method. The basic experiment was replicated 1000 times for each combination of the sample sizes and the underlying distributions of the vectors. The nominal level was 0.05. The proportion of times that the values of the statistics exceeded the critical values was recorded as the empirical power.

Table 3.1   Power of the tests with $n = 20$

		b	0.00	0.25	0.50	0.75	1.00	1.25
$d = 2$	T_{n1}		0.046	0.201	0.332	0.584	0.776	0.870
	T_{n2}		0.043	0.223	0.391	0.578	0.630	0.663
$d = 4$	T_{n1}		0.045	0.181	0.315	0.577	0.679	0.861
	T_{n2}		0.040	0.238	0.402	0.579	0.633	0.643
$d = 6$	T_{n1}		0.053	0.195	0.345	0.581	0.667	0.860
	T_{n2}		0.038	0.251	0.407	0.576	0.616	0.654

Table 3.2   Power of the tests with $n = 50$

	b	0.00	0.25	0.50	0.75	1.00	1.25
$d = 2$	T_{n1}	0.046	0.261	0.392	0.664	0.874	0.950
	T_{n2}	0.046	0.283	0.455	0.648	0.797	0.853
$d = 4$	T_{n1}	0.045	0.281	0.385	0.637	0.881	0.957
	T_{n2}	0.046	0.288	0.462	0.635	0.831	0.850
$d = 6$	T_{n1}	0.053	0.295	0.395	0.640	0.866	0.960
	T_{n2}	0.043	0.311	0.487	0.641	0.818	0.846

Looking at Table 3.1 with $n = 20$, we see that, under the null hypothesis, that is $b = 0.00$, the size of the test T_{n1} is close to the nominal level and T_{n2} is somewhat conservative. But it gets better with the increasing of sample size to $n = 50$. See Table 3.2. Under the alternatives, namely $b \neq 0.00$, when b is small, it seems that T_{n2} with the estimated center and shape matrix would be more sensitive to alternatives than is T_{n1}. See the cases with $b = 0.25, 0.50$. With larger b, the situation is reversed. Since the location is not needed to estimate, NMCT with T_{n1} should simulate the null distribution of the test statistic better. Hence it is understandable that T_{n1} has better performance at maintaining the significance level than does T_{n2}. When $n = 50$, the margin is small. It seems that $n = 50$ can be regarded a large size of sample. See Table 3.2. Also when $n = 50$ the power of the tests is higher than that with $n = 20$. Furthermore, the power performance of the tests are less affected by the dimension of variable.

3.4 Appendix: Proofs of Theorems

In this section, we only present the proofs of theorems about elliptical symmetry testing. Similar arguments can be applied to prove the theorems for reflection symmetry testing; we omit them here.

Proof of Theorem 3.2.1 Ghosh and Ruymgaart (1992) have proved that, when the center and the shape matrix are given, the process V_n converges in distribution to V_1 with the covariance kernel in (3.2.4). When the shape matrix is replaced by the sample covariance matrix $\hat{\Sigma}$, applying the triangle identity, we have

$$\sqrt{n}P_n(\sin(ta^\tau \hat{A}(X - \mu)))$$
$$= \sqrt{n}P_n(\sin(ta^\tau A(X - \mu)))\cos(ta^\tau(\hat{A} - A)(X - \mu)))$$
$$+\sqrt{n}(P_n(\cos(ta^\tau A(X - \mu)))\sin(ta^\tau(\hat{A} - A)(X - \mu)))$$
$$=: I_{n1}(t,a) + I_{n2}(t,a).$$

It is well-known that by the conditions $\max_{1 \le j \le n} ||X_j - \mu||/n^{1/4} \to 0$, a.s., $\sqrt{n}(\hat{A}A^{-1} - I_d) = O_p(1)$, and $E\left(A(X - \mu)\cos(ta^\tau A(X - \mu))\right) = 0$ which is implied by the spherical symmetry of $A(X - \mu)$, we then easily derive that, uniformly over $(t, a) \in I \times S^d$,

$$I_{n1}(t, a) = \sqrt{n}P_n(\sin(ta^\tau A(X - \mu))) + O_p(1/\sqrt{n}),$$
$$I_{n2}(t, a) = ta^\tau \sqrt{n}(\hat{A}A^{-1} - I_d)(P_n(A(X - \mu)\cos(ta^\tau A(X - \mu)))$$
$$= o_p(1).$$

This implies that V_n with the sample covariance matrix $\hat{\Sigma}$ is asymptotically equivalent to that with Σ. Conclusion 1) is proved. For conclusion 2), the argument is analogous since we can derive that

$$\sqrt{n}\,P_n(\sin(ta^\tau \hat{A}(X - \hat{\mu}))) = \sqrt{n}\,P_n(\sin(ta^\tau \hat{A}(X - \mu)))\cos(ta^\tau \hat{A}(\hat{\mu} - \mu)))$$
$$-\sqrt{n}\,(P_n(\cos(ta^\tau \hat{A}(X - \mu)))\sin(ta^\tau \hat{A}(\hat{\mu} - \mu)))$$
$$= \sqrt{n}\,P_n(\sin(ta^\tau \hat{A}(X - \mu)))$$
$$-\sqrt{n}\,ta^\tau \hat{A}(\hat{\mu} - \mu))E(\cos(ta^\tau \hat{A}(X - \mu))) + o_p(1).$$

The proof of Theorem 3.2.1 is completed.   □

Proof of Theorem 3.2.2. Consider the case of $\hat{\mu} = \mu$ first. Assume no loss of generality that the center $\mu = 0$ and the covariance matrix of X_{in} is $\Sigma_n = (A_n)^{-2}$. Note that Σ_n converges to the covariance matrix of the variable Z, Σ say. Applying the Taylor expansion to the *sine* function, for any $(t, a) \in I \times S^d$,

$$\sqrt{n}P_n\{\sin(ta^\tau A_n(Z + \frac{Y}{n^{1/(2l)}}))\}$$

$$= \sqrt{n}P_n\{\sin(ta^\tau A_n Z)\} + \sum_{i=1}^{l-1} \frac{1}{i!}n^{-i/(2l)}\sqrt{n}P_n\{(ta^\tau A_n Y)^i \sin^{(i)}(ta^\tau A_n Z)\}$$

$$+ \frac{1}{l!n}\sum_{j=1}^{n}\{(ta^\tau A_n Y_j)^l \sin^{(l)}(ta^\tau A_n(Z_j + \frac{(t^\tau Y_j)^*}{n^{1/(2l)}}))) - \sin^l(ta^\tau A_n Z_j))\}$$

$$+ \frac{1}{l!}P_n\{(ta^\tau A_n Y)^l \sin^{(l)}(ta^\tau A_n Z)\}, \tag{3.4.1}$$

where $(ta^\tau A_n Y_j)^*$ is a value between 0 and $ta^\tau A_n Y_j$. We need to show that the second and third summands on RHS of (3.4.1) tend to zero in probability as $n \to \infty$, and the fourth summand converges in probability to $E\{(ta^\tau AY)^l)\sin^{(l)}(ta^\tau AZ)\}$. The convergence of the fourth term is obvious. Noticing that $E\{(ta^\tau AY)^i)\sin^{(i)}(ta^\tau AZ)\} = 0$ for $1 \le i \le l-1$, and a similar argument used in the proof of Theorem 3.2.1 can be applied. The proof for V_n with the known center and then for T_n is finished.

For V_n with an estimated covariance matrix, we note that

$$\max_{1\leq j\leq n} ||Y_j||/n^{1/(2l)} \to 0, \quad a.s.$$

$$\sqrt{n}(\hat{A}_n - A_n) = O_p(1) \quad \text{and} \quad A_n - A = o(1).$$

Furthermore,

$$\sup_{(t,a)\in I\times S^d} \left| P_n\Big(\sin(ta^\tau \hat{A}_n(Z - E(Z)) + (Y - E(Y)/n^{1/(2l)})) \right.$$

$$\left. - \sin(ta^\tau \hat{A}_n(Z - E(Z)))\Big) \right|$$

$$\leq cP_n||\hat{A}_n(Y - E(Y))||/n^{1/(2l)} = O(n^{-1/(2l)}) \quad a.s.,$$

and

$$\sup_{(t,a)\in I\times S^d} |1 - \cos(ta^\tau P_n(\hat{A}_n(Z - E(Z)) + \hat{A}_n(Y - E(Y)/n^{1/(2l)})))|$$

$$\leq c(||P_n\hat{A}_n(Z - E(Z))||^2 + ||P_n\hat{A}_n(Y - E(Y))||^2/n^{1/l} = O_p(n^{-1}).$$

Similar argument used in the proof of Theorem 3.2.1 can be applied again. The details are omitted. From the convergence of V_n we immediately derive the convergence of T_n in (3.2.8).

For the case of $\hat{\mu} = \bar{X}$, we further note that

$$\sup_{(t,a)\in I\times S^d} \left| \sqrt{n}\Big(\sin(ta^\tau P_n\hat{A}_n((Z - E(Z)) + (Y - E(Y)/n^{1/(2l)}))) \right.$$

$$\left. - \sin(ta^\tau P_n\hat{A}_n(Z - E(Z)))\Big) \right|$$

$$\leq c\sqrt{n}||\hat{A}_n(P_nY - E(Y))||/n^{1/(2l)} = O_p(n^{-1/(2l)}).$$

Based on the above inequalities and the triangle identity, it is easy to see that

$$\sqrt{n}P_n(\sin(ta^\tau \hat{A}_n(Z + Y/n^{1/(2l)}) - (\bar{Z} + \bar{Y}/n^{1/(2l)}))))$$

$$= \sqrt{n}P_n(\sin(ta^\tau A_n(Z + Y/n^{1/(2l)}) - (E(Z) + E(Y)/n^{1/(2l)}))))$$

$$-\sqrt{n}P_n(\cos(ta^\tau A_n(Z - E(Z)))\sin(ta^\tau P_n A_n(Z - E(Z))) + O_p(n^{-1/(2l)})$$

$$= \sqrt{n}P_n(\sin(ta^\tau A(Z - E(Z))))$$

$$+\frac{1}{l!}E\{(ta^\tau A(Y - E(Y)))^l \sin^{(l)}(ta^\tau A(Z - E(Z)))\}$$

$$-\sqrt{n}\sin(ta^\tau P_n A(Z - E(Z)))E(\cos(ta^\tau A(Z - E(Z))) + o_p(1)$$

$$\Longrightarrow V_2(t, a) + (1/l)B_l(t, a). \tag{3.4.2}$$

It implies the convergence of T_n in (3.2.9). The proof is completed. $\square$

Proof of Theorem 3.3.1. We only need to show the convergence of the processes, which implies the convergence of the test statistics. First we show that $\{V_{n1}(\mathbf{U}_n, \mathbf{X}_n, t, a) : (t, a) \in I \times S^d\}$ given $\mathbf{X}_n$ converges almost surely to the process $\{V_1(t) : (t, a) \in I \times S^d\}$ which is the limit of V_n with the known center. The argument of the proof will be applicable for showing the convergence of the process $V_{n2}(\mathbf{U}_n)$.

For simplicity of notation, write X_j for $A(X_j - \mu)$. Define sets

$$D_1 = \{\lim_{n \to \infty} \frac{1}{n} \sum_{j=1}^{n} \|X_j\|^2 = E\|X\|^2\},$$

$$D_2 = \{\lim_{n \to \infty} \sup_{(t,a),(s,b)} \left| \frac{1}{n} \sum_{j=1}^{n} (\sin(ta^\tau X_j) \sin(sb^\tau X_j)) \right.$$

$$\left. - E(\sin(ta^\tau X) \sin(sb^\tau X)) \right| = 0\}$$

and $D = D_1 \cap D_2$. By the Lipschitz continuity of the $sine$ function and the Glivenko-Cantelli theorem for the general class of functions (e.g. Pollard (1984), Theorem II 24, pp. 25), it is clear that D is a subset of sample space with probability measure one.

We assume without further mentioning that $\{X_1, \cdots, X_n, \cdots\} \in D$ in the following.

For the convergence of the empirical process defined in the theorem, all we need to do is to prove *fidis convergence* and *uniform tightness*. The proof of the *fidis convergence* is standard, so we only describe an outline. For any integer k, $(t_1, a_1) \cdots (t_k, a_k) \in I \times S^d$. Let

$$V^{(k)} = \left(\mathrm{cov}(\sin(t_i a_i^\tau x), \sin(t_l a_l^\tau x)) \right)_{1 \le i, l \le k}.$$

It needs to be shown that

$$V_{n1}^{(k)} = \{V_{n1}(\mathbf{U}_n, \mathbf{X}_n, t_i, a_i) : i = 1, \cdots, k\} \Longrightarrow N(0, V^{(k)}).$$

It suffices to show that for any unit k-dimensional vector γ

$$\gamma^\tau V_{n1}^{(k)} \Longrightarrow N(0, \gamma^\tau V^{(k)} \gamma). \qquad (3.4.3)$$

Note that the variance of LHS in (3.4.3), as follows, converges in probability to $\gamma^\tau V^{(k)} \gamma$

$$\gamma^\tau \left(\widehat{Cov}_{i,l} \right)_{1 \le i, l \le k} \gamma,$$

with $\widehat{Cov}_{i,l} = \frac{1}{n} \sum_{j=1}^{n} E(\sin(t_i a_i^\tau u \|X_j\|) \sin(t_l a_l^\tau u \|X_j\|))$ where the expectation is taken over u. Hence if $\gamma^\tau V^{(k)} \gamma = 0$, (3.4.3) is trivial. Assume $\gamma^\tau V^{(k)} \gamma > 0$. Invoking the boundedness of the $sine$ function and the Lindeberg condition for central limit theorems,

$$\gamma^\tau V_{n1}^{(k)} / \sqrt{\gamma^\tau V^{(k)} \gamma} \longrightarrow N(0, 1).$$

That is, (3.4.3) holds, and the *fidis convergence* is then proved.

We now turn to prove the *uniform tightness* of the process. All we need to do is to show that for any $\eta > 0$ and $\epsilon > 0$, there exists an $\delta > 0$ for which

$$\limsup_{n\to\infty} P\left\{ \sup_{[\delta]} |V_{n1}(\mathbf{U}_n, \mathbf{X}_n, t, a) - V_{n1}(\mathbf{U}_n, \mathbf{X}_n, s, b)| > 2\eta \Big| \|\mathbf{X}_n\| \right\} < \epsilon$$

(3.4.4)

where $[\delta] = \{((t,a),(s,b)) : \|ta - sb\| \le \delta\}$. Since the limiting properties are investigated with $n \to \infty$, n is always considered to be large enough below, which simplifies some arguments of the proof.

It is easy to show that if a d-dimensional vector u is uniformly distributed on S^d, then u can be expressed as $e \cdot u^*$ where $e = \pm 1$ with probability one half, u^* has the same distribution as u and e and u^* are independent. The justification can be done by noting that for this e independent of u, eu has the same distribution as u and e is also independent of eu. From which, the LHS of (3.4.4) can be written as

$$P\left\{ \sup_{[\delta]} \sqrt{n}|P_n(\sin(ta^\tau e \bullet u^*\|X\|) - \sin(sb^\tau e \bullet u^*\|X\|))| > \eta \Big| \|\mathbf{X}_n\| \right\}$$

$$= P\left\{ \sup_{[\delta]} \sqrt{n}|P_n^\circ(\sin(ta^\tau u^*\|X\|) - \sin(sb^\tau u^*\|X\|))| > \eta \Big| \|\mathbf{X}_n\| \right\} \quad (3.4.5)$$

where P_n° is the signed measure that places mass e_i/n at $u_i\|X_i\|$, which is analogous to that of Pollard (1984, p.14).

We now consider conditional probability given $\mathbf{U}_n^* = (u_1^*, \cdots, u_n^*)$ and $\|\mathbf{X}_n\|$. Combining (3.4.5) with the following inequality

$$|\sin(ta^\tau u^*\|X\|) - \sin(sb^\tau u^*\|X\|)| \le \|ta - sb\| \|X\|,$$

the Hoeffding inequality implies that

$$P\left\{ \sqrt{n}|(P_n^\circ(\sin(ta^\tau u^*\|X\|) - \sin(sb^\tau u^*\|X\|))) > \eta c\|ta - sb\| \Big| \|\mathbf{X}_n\|, \mathbf{U}_n^* \right\}$$

$$\le 2\exp(-\eta^2/32).$$

In order to apply the chaining lemma (e.g. Pollard (1984), p.144), we need to check, together with the above inequality, the covering integral

$$J_2(\delta, \|\cdot\|, I \times S^d) = \int_0^\delta \{2\log\{(N_2(u, \|\cdot\|, I \times S^d))^2/u\}\}^{1/2} du \quad (3.4.6)$$

is finite for small $\delta > 0$, where $\|\cdot\|$ is the Euclidean norm in R^d and the covering number $N_2(u, \|\cdot\|, I \times S^d)$ is the smallest l for which there exist l points $t_1, \cdots, t_l$ with $\min_{1\le i\le l} \|ta - t_i a_i\| \le u$ for every $(t,a) \in I \times S^d$. It is clear that

$$N_2(u/c, \| \cdot \|, I \times S^d) \leq cu^{-d}.$$

Consequently, for small $\delta > 0$,

$$J_2(\delta, \| \cdot \|, I \times S^d) \leq c \int_0^\delta (-\log u)^{1/2} du \leq c\delta \log \delta \leq c\delta^{1/2}.$$

Therefore (3.4.6) holds. Applying now the chaining lemma, there exists a countable dense subset $[\delta]^*$ of $[\delta]$ such that

$$P\Big\{ \sup_{[\delta]^*} \sqrt{n} |(P_n^\circ(\sin(ta^\tau u^* \|X\|) - \sin(sb^\tau u^* \|X\|))|$$

$$> 26cJ_2(\delta, \| \cdot \|, I \times S^d)\Big| \|\mathbf{X}_n\|, \mathbf{U}_n^*\Big\}$$

$$\leq 2c\delta.$$

The countable dense subset $[\delta]^*$ can be replaced by $[\delta]$ itself because

$$\sqrt{n} P_n^\circ \{\sin(ta^\tau u \|X\|) - \sin(sb^\tau u \|X\|)\}$$

is a continuous function with respect to ta and sb for each fixed $\|\mathbf{X}_n\|$. Hence, choosing properly small δ, and integrating out over $\mathbf{U}_n^*$, the uniform tightness in (3.4.4) is proved. Therefore, the convergence of the process is proved. Then the convergence of $T_{n1}(\mathbf{U}_n)$ follows. The convergence of the process $V_{n2}(\mathbf{U}_n)$ can be proved by following the above argument and noticing $\hat{A} - A = O_p(1/\sqrt{n})$ and $\hat{\mu} - \mu = O_p(1/\sqrt{n})$. The limit of $V_{n2}(\mathbf{U}_n)$ is V_2, the limit of its unconditional counterparts. The asymptotic validity of T_{n2} then follows. The proof of Theorem 3.3.1 is finished. $\qquad\square$

4

A Test of Dimension-Reduction Type for Regressions

4.1 Introduction

Parametric models describe the impact of the covariate X on the response Y in a concise way. They are easy to use. But since there are usually several competing models to entertain, model checking becomes an important issue.

Suppose that $\{(x_1, y_1), \cdots, (x_n, y_n)\}$ are *i.i.d.* observations satisfying the following relation:

$$y_i = \phi(x_i) + \varepsilon_i \qquad i = 1, \cdots, n, \tag{4.1.1}$$

where y_i is one-dimensional, $x_i = (x_i^{(1)}, \cdots, x_i^{(d)})'$ is the d-dimensional column vector and ε_i is independent of x_i. We want to test

$$H_0 : \phi(x) = \phi_0(\cdot, \beta) \quad \text{for some } \beta, \tag{4.1.2}$$

where $\phi_0(\cdot, \cdot)$ is a specified function.

For this testing problem, there are a number of non-parametric approaches available in the literature. One approach to constructing a test statistic is through a suitable estimate of $\phi(\cdot) - \phi_0(\cdot, \beta)$. The local smoothing for estimating ϕ is often employed. See Härdle and Mammen (1993). The success of this locally smoothing method hinges on the presence of sufficiently many data points to provide adequate local information. For one-dimensional cases, many smoothing techniques are available and the tests obtained have good performance; the book by Hart (1997) gives an extensive overview and useful references. As the dimension of the covariate gets higher, however, the total number of observations needed for local smoothing escalates exponentially. Another approach is to resort to the ordinary residuals $\hat{\varepsilon}_i = y_i - \phi_0(x_i, \beta_n)$, where β_n is an estimate of β. Globally smoothing method includes CUSUM tests (Buckley, 1991; Stute, 1997; Stute, González Manteiga and Presedo Quindimil, 1998); the innovation transformation based tests (Stute, Thies and Zhu, 1998; Stute and Zhu, 2002); and the score type tests (Stute and Zhu, 2005).

For a practical point of view, however, these two testing approaches suffer from the lack of flexibility in detecting subtle dependence patterns between the residuals and the covariates. As a remedy, practitioners often rely on residual plots, i.e. plots of residuals against fitted values or a selected number of covariates for model checking. But this poses a problem when the number of covariates is large especially when one wants to include all possible linear combinations of covariates.

In this chapter, we recommend a dimension-reduction type test for seeking a good projection direction for plotting and constructing a test statistic. Most of materials are from Zhu (2003).

For any fixed t, consider

$$I_n(t) = \frac{1}{\sqrt{n}} \sum_{j=1}^{n} \hat{\Sigma}^{-1/2}(x_j - \bar{x}) I(\hat{\varepsilon}_j \leq t), \qquad (4.1.3)$$

where $I(\hat{\varepsilon}_j \leq t) = 1$ if $\hat{\varepsilon}_j \leq t$, and $= 0$, otherwise; $\hat{\Sigma}$ is the sample covariance matrix of x_i's. For any $a \in S^d = \{a : ||a|| = 1\}$, define

$$T_n(a) = a^\tau \left[\frac{1}{n} \sum_{i=1}^{n} (I_n(x_i) I_n^\tau(x_i)) \right] a. \qquad (4.1.4)$$

One test statistic is defined by

$$T_n := \sup_{a \in S^d} T_n(a). \qquad (4.1.5)$$

In this chapter, the estimate β_n of β is given by the least squares method, that is,

$$\beta_n = \arg\min_{\beta} \sum_{j=1}^{n} (y_j - \phi_0(x_j, \beta))^2.$$

The maximizer $\mathbf{a}$ of $T_n(a)$ over $a \in S^d$ will be used as the projection direction to plot the residuals. Note that T_n and $\mathbf{a}$ are simply the largest eigenvalue and the associated eigenvector of the matrix $\left[\frac{1}{n} \sum_{i=1}^{n} (I_n(x_i) I_n^\tau(x_i)) \right]$, therefore the implementation is easy.

The motivation is quite simple. If the model is correct, $e = y - \phi_0(x, \beta)$ is independent of x. Under the null hypothesis H_0

$$E(\Sigma^{-1/2}(X - EX)|e) = 0$$

where Σ is the covariance matrix of X. This is equivalent to

$$I(t) = E[\Sigma^{-1/2}(X - E(X)) I(e \leq t)] = 0 \quad \text{for all } t \in R^1.$$

Consequently, for any $a \in S^d$

$$T(a) := a^\tau \left[\int (I(t))(I(t))^\tau dF_e(t) \right] a = 0,$$

where F_e is the distribution of e. Then the test statistic $T_n = \sup_a T_n(a)$ is the empirical version of $\sup_a T(a)$. The null hypothesis H_0 is rejected for the large values of T_n.

Note that the test T_n does not involve the local smoothing, and the determination of the projection direction is easily done. The dimensionality problem may largely be avoided. The next section contains the limit behavior of the test statistic. For computing the p-value, the consistency of bootstrap approximations and NMCT is discussed in Section 4.3. A simulation study on the power performance and the comparison between the bootstrap and NMCT are reported in Section 4.4. The residual plot is also presented in this section. Section 4.5 contains some further remarks. Proofs of the theorems in Sections 4.2 and 4.3 are postponed to Section 4.6.

4.2 The Limit Behavior of Test Statistic

Before stating the theorem, we present the linear representation of the least squares estimate β_n first. Note that under certain regularity conditions, β_n can be written as

$$\beta_n - \beta = \frac{1}{n} \sum_{j=1}^n L(x_j, \beta)\varepsilon_j + o_p(1/\sqrt{n}),$$

where, letting ϕ_0' be the derivative of ϕ_0 at β, $L(X, \beta) = (E[(\phi_0')(\phi_0')^\tau])^{-1} \times \phi_0'(X, \beta)$. Especially, when ϕ_0 is the linear function

$$\beta_n = S_n^{-1} X_n Y_n$$

with $X_n = \{x_1 - \bar{x}, \cdots, x_n - \bar{x}\}, Y_n = \{y_1 - \bar{y}, \cdots, y_n - \bar{y}\}$ and $S_n = (X_n X_n^\tau)$.

We now state the asymptotic result of T_n. Let $V_1(X) = \left(E[\Sigma^{-1/2}(X - E(X))(\phi_0'(X, \beta))^\tau] \right) L(X, \beta)$.

THEOREM 4.2.1 *Assume that the density function f_ε of ε exists, the derivative $\phi_0'(X, \beta)$ of $\phi_0(X, \beta)$ at β is continuous and has $(2 + \delta)$-th moment for some $\delta > 0$, and both the covariance matrix of $\phi_0'(X, \beta)$ and Σ are positive definite. Then under H_0*

$$I_n(t) = \frac{1}{\sqrt{n}} \sum_{j=1}^n \Sigma^{-1/2}(x_j - E(X)) \left(I(\varepsilon_j \leq t) - F_\varepsilon(t) \right)$$

$$+ f_\varepsilon(t) \left(E[\Sigma^{-1/2}(X - E(X))(\phi_0'(X, \beta))^\tau] \right) \frac{1}{\sqrt{n}} \sum_{j=1}^n L(x_j, \beta)\varepsilon_j + o_p(1),$$

and then T_n converges in distribution to a vector of Gaussian process

$$I = B - f_\varepsilon \cdot N \qquad (4.2.1)$$

in the Skorohod space $D^d[-\infty, \infty]$, where B is a vector of Gaussian processes $(B_1, \cdots B_d)^\tau$ with the covariance function $cov(B_i(t), B_i(s)) = F_\varepsilon(min(t, s)) - F_\varepsilon(t)F_\varepsilon(s)$, $F_\varepsilon(t)$ and $f_\varepsilon(t)$ are respectively the distribution and density functions of ε, and N is a random vector with a normal distribution $N(0, \sigma^2 V)$ with $V = E(V_1 V_1^\tau)$. The covariance function of each component of I, $I^{(i)}$ say, is that for $s \leq t$

$$
\begin{aligned}
K^{(i)}(s, t) = &F_\varepsilon(s) - F_\varepsilon(s)F_\varepsilon(t) + f_\varepsilon(s)f_\varepsilon(t)E(V_1^{(i)})^2 \\
&- f_\varepsilon(s) \int \varepsilon I(\varepsilon \leq t) dF_\varepsilon E(V_1^{(i)}(\Sigma^{-1/2}(X - E(X)))^i) \\
&- f_\varepsilon(t) \int \varepsilon I(\varepsilon \leq s) dF_\varepsilon E(V_1^{(i)}(\Sigma^{-1/2}(X - E(X)))^i), \qquad (4.2.2)
\end{aligned}
$$

where $(\Sigma^{-1/2}(X - E(X)))^i$ is the i-th component of $\Sigma^{-1/2}(X - E(X))$. The process convergence implies that T_n converges in distribution to $T = \sup_a a^\tau \left(\int (I(t)I(t)^\tau) dF_\varepsilon(t) \right) a$.

But the result of Theorem 4.2.1 does not help in determining the p-values because the distribution of T is intractable. Monte carlo approximations will be helpful.

4.3 Monte Carlo Approximations

We first consider bootstrap approximations. The basic bootstrap procedure for our setup is as follows: Let (x_i^*, y_i^*), $i = 1, \cdots, n$ be a reference dataset, and let β_n^* be the least squares estimator computed from this sample; a conditional counterpart of I_n given $\{(x_1, y_1), \cdots, (x_n, y_n)\}$ is defined by

$$I_n^*(t) = n^{-\frac{1}{2}} \sum_{j=1}^n (\hat{\Sigma}^*)^{-1/2}(x_j^* - \bar{x}^*)I(\hat{\varepsilon}_j^* \leq t), \qquad (4.3.1)$$

where $\hat{\varepsilon}_j^*$'s are the residuals based on (x_i^*, y_i^*)'s, that is, $\hat{\varepsilon}_j^* = y_j^* - \phi_0(x_j^*, \beta_n^*)$ (or $\hat{\varepsilon}_j^* = y_j^* - (\beta_n^*)^\tau x_j^*$ when ϕ_0 is linear) and $(\hat{\Sigma}^*)^{-1/2}$ is the covariance matrix of x_i^*'s. The conditional counterpart of T_n will be

$$T_n^* = \sup_a a^\tau \left[\int (I_n^*(t))(I_n^*(t))^\tau dF_n^*(t) \right] a, \qquad (4.3.2)$$

where F_n^* is the empirical distribution based on ε_i^*, $i = 1, \cdots, n$. For computing the p-values, we generate m sets of data $\{(x_j^*, y_j^*), j = 1, \cdots, n\}^{(i)}$, $i = 1, \cdots, m$, then compute m values of T_n^*. The p-value is estimated by

$\hat{p} = k/m$ where k is the number of T_n^*'s which is larger than or equal to T_n. For the nominal level α, when $\hat{p} \leq \alpha$, the null hypothesis is rejected.

There are three Monte Carlo approximations to use: Wild bootstrap, NMCT and the classical bootstrap. We describe their algorithms below.

Option 1. Wild Bootstrap.   Define

$$x_i^* = x_i \text{ and } y_i^* = \phi_0(x_i, \beta_n) + \varepsilon_i^*$$

where ε_i^* are defined as

$$\varepsilon_i^* = w_i^* \hat{\varepsilon}_i$$

and w_i^* are *i.i.d.* artificial bounded variables with

$$E(w_i^*) = 0, \text{ Var}(w_i^*) = 1 \text{ and } \quad E|w^*|^3 < \infty. \tag{4.3.3}$$

The bootstrap residuals $\hat{\varepsilon}_i^* = y_i^* - \phi_0(x_i, \beta_n^*)$ which are used to construct the bootstrap process I_{n1}^* and then the test statistic T_{n1}^*, like those in (4.3.1) and (4.3.2).

Option 2. NMCT.   Let

$$x_i^* = e_i(x_i - \bar{x}) \text{ and } \hat{\varepsilon}_i^* = \hat{\varepsilon}_i - (\frac{1}{n}\sum_{j=1}^{n} e_j L(x_j, \beta_n)\hat{\varepsilon}_j)e_i\phi_0'(x_i, \beta_n), \tag{4.3.4}$$

where the weight variables e_i are the same as those in the wild bootstrap and $L(\cdot, \cdot)$ is defined in the estimate β_n. When the model is linear, (4.3.4) reduces to $x_j^* = e_j(x_j - \bar{x})$ when e_j is replaced by w_j^* and

$$\hat{\varepsilon}_i^* = \hat{\varepsilon}_i - S_n^{-1}(\frac{1}{n}\sum_{j=1}^{n} x_j^*\hat{\varepsilon}_j)x_i^* =: \hat{\varepsilon}_i - (\beta_n^*)^\top x_i^*. \tag{4.3.5}$$

The NMCT process and the resulting statistic can then be created, say I_{n2}^* and T_{n2}^*.

Option 3. The Classical Bootstrap.   Draw the independent bootstrap data from the residuals $\hat{\varepsilon}_i$, say $e_1^*, \cdots, e_n^*$. Define

$$x_i^* = x_i \text{ and } y_i^* = \phi_0(x_j, \beta_n) + e_i^*.$$

The bootstrap residuals $e_i^* = y_i^* - \phi_0(x_j, \beta_n^*)$, which are used to define the bootstrap process and the test statistic like those in (4.3.1) and (4.3.2), say I_{n3}^* and T_{n3}^*.

As shown in the literature, Wild Bootstrap is a useful approximation in model checking for regression, Stute, González Manteiga and Presedo

Quindimil (1998) formally proved that, when an test based on the residual marked empirical process is applied, the classical bootstrap fails to work but the wild bootstrap works. It is also true when a test based on the squared distance between the parametric and nonparametric fit of data. See Härdle and Mammen (1993). Interestingly the situation in our case is reversed, that is, the classical bootstrap is consistent while the wild bootstrap is inconsistent. The consistency of the NMCT is also true. The following theorem states the consistency of the classical bootstrap and NMCT approximations.

THEOREM 4.3.1 *Under H_0 and the assumptions in Theorem 2.1, we have that with probability one both I_{n2}^* and I_{n3}^* converge weakly to I^* in the Skorohod space $D^d[-\infty, \infty]$. where I^* has the same distribution as I defined in Theorem 4.2.1*

The inconsistency of the wild bootstrap is as follows.

THEOREM 4.3.2 *In addition to the assumptions in Theorem 2.1 assume that $w^* = \pm 1$ with one a half probability and the density of ε is symmetric about the origin. Under H_0, I_{n1}^* does not converge in distribution to $B - f_\varepsilon \cdot N$.*

4.4 Simulation Study

4.4.1 Power Study

In order to demonstrate the performance of the proposed test procedures, small-sample simulation experiments were performed. Since the wild bootstrap is not consistent with our test statistic, we made a comparison among Stute , González Manteiga and Presedo Quindimil's (1998) test (T_S^*), the NMCT (T_{n2}^*)(Option 2) and the classical bootstrap test (T_{n3}^*)(Option 3). The model was

$$y = a^\tau x + b(c^\tau x)^2 + \varepsilon \qquad (4.4.1)$$

where x is d-dimensional covariate. The dimension $d = 3, 6$. When $d = 3$, $a = [1, 1, 2]^\tau$ and $c = [2, 1, 1]^\tau$ and for 6-dimensional case $a = [1, 2, 3, 4, 5, 6]^\tau$ and $c = [6, 5, 4, 3, 2, 1]^\tau$. Furthermore, let $b = 0.00, 0.3, 0.7, 1.00, 1.50$ and 2.00 to provide evidence on the power performance of the test under local alternatives. $b = 0.00$ corresponds to the null hypothesis H_0. The sample size $n = 25$ and 50. The nominal level was 0.05. In each of the 1000 replicates, 1000 bootstrap samples were drawn.

Figure 4.1 presents the power of the tests. From it, we have three findings on the comparison. First, looking at Figure 4.1(1) and 4.1(2), we find that with increasing the size of sample, both T_{n2}^* and T_{n3}^* improve their performance more quickly than T_S^*. Figure 4.1(3) and 1(4) also indicate this tendency. Second, in the 6-dimensional cases, T_S^* has much higher power but cannot maintain the size of the test. The size are $0.09(d = 6, n = 25)$ and $0.083(d = $

$6, n = 50$). Both T_{n2}^* and T_{n3}^* are a bit conservative. Third, T_{n2}^* and T_{n3}^* are competitive with each other. As T_{n2}^* is computationally more efficient, it may be worthwhile to recommend.

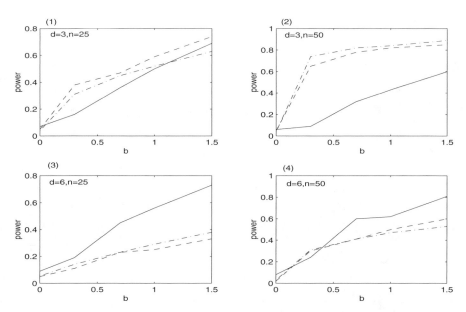

Fig. 4.1. Plots (1) and (2) are with 3-dimensional covariate and plots (3) and (4) are for 6-dimensional case. The solid, dot dashed and dashed lines are respectively for the power of T_S^*, T_{n2}^* and T_{n3}^*.

4.4.2 Residual Plots

In addition to the formal test, we also consider the plots of $\hat{\varepsilon}_i$ against the projected covariate $\alpha^\tau x_i$ along the direction α selected by (4.1.5). We use model (4.4.1) with $b = 0$ and $b = 1$ to generate $n = 50$ data points. $b = 0$ and $b = 1$ correspond respectively to linear and nonlinear models. Figure 4.2 presents the plots of the residuals versus the projected covariates for linear and nonlinear model respectively when the models are fitted linearly. Plots (1) and (3) in Figure 4.2 show that there is no clear indication of relationship while plots (2) and (4) show some relationship.

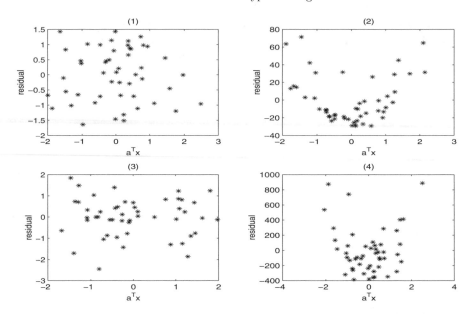

Fig. 4.2. The residual plots against $\alpha^\tau x$ where α is determined by (4.1.5). Plots (1) and (2) are for the 3-dimensional cases with $b = 0$ and $b = 1$ and plots (3) and (4) are for 6-dimensional cases with $b = 0$ and $b = 1$.

4.4.3 A Real Example

We now analyze a real data set. Consider the 1984 Olympic records data of men on various track events as reported by Dawkins (1989). Principal component analysis has been applied to study the athletic excellence of a given nation and the relative strength of the nation at the various running distances. For the 55 countries the winning times for men's running events of 100, 200, 400, 800, 1,500, 5,000 and 10,000 meters and Marathon are reported in Dawkins. It is of interest to study whether a nation whose performance is better in running long distances may also have greater strength at short running distances. It may be more reasonable to use the speed rather than the winning time for the study as did Naik and Khattree (1996). Convert the winning time to the speed defined as $x_1, \ldots, x_8$. From principal component analysis we may regard 100, 200 and 400 meters as short running distances and 1,500 meters and longer as long running distances. A linear model was fitted by considering the speed of the 100 meters running event (x_1) as the response and the speed of the 1,500, 5,000 and 10,000 meters and Marathon running events $(x_5, \ldots, x_8)$ as covariates. The p-values of T_S^*, T_{n2}^* and T_{n3}^* are 0.02, 0.08 and 0.01. We may have to reject the null hypothesis that a linear relationship between the speed of the 100 meters running event and the long distance running events is assumed. Looking at Figure 3 which presents the plots of the residuals versus $\alpha^\tau x$ we find that Figure 4.3(1) shows some relationship. But after removing the point of Cook Islands, no clear indication of relationship is presented. Using T_S^*, T_{n2}^* and T_{n3}^* again for the data except that of Cook Islands, the p-values are 0.57, 0.64 and 0.06 respectively. Therefore, the linear model may be tenable. We may regard the point of Cook Islands as an outlier.

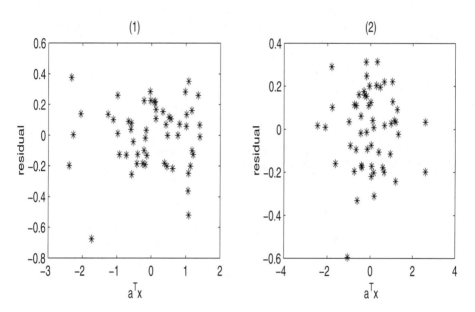

Fig. 4.3. (1) and (2) are the residual plots against $\alpha^\tau x$ where α is determined by (4.1.5). Plots are for the countries with and without Cook Islands respectively.

4.5 Concluding Remarks

In this chapter, we recommend a dimension-reduction approach for model checking for regression models. The formal test and the residual plots can be constructed in terms of the projected covariate. The implementation is very easy. A negative aspect of our approach is that the test may highly rely on the assumption of independence between the covariate X and the error ε. This suggests that the test may be difficult to apply to the case where only $E(\varepsilon|x) = 0$ is assumed. Furthermore, an interesting finding is that the wild bootstrap is inconsistent, but the NMCT and the classical bootstrap work. This means that the use of bootstrap approximations needs to be delicate.

4.6 Proofs

For simplifying the presentation of proofs, throughout this section we assume no loss of generality that X is scalar and $\phi_0(X, \beta)$ is a linear function, $\beta^\tau X$. We can assume so because asymptotically $\phi_0(X, \beta_n) - \phi_0(X, \beta) = (\beta_n - \beta)^\tau \phi_0'(X, \beta)$ and $\beta_n - \beta$ has an asymptotically linear representation like that in the linear model case, hence the proof for the general ϕ_0 is almost the same as that for linear function. We also assume that Σ is the identity one and $\hat{\Sigma}$ is replaced by the identity matrix. This replacement does not affect the asymptotic results at all.

We first present a lemma. The idea of proving it will serve as an useful tool for all proofs of the theorems.

LEMMA 4.6.1 *Assume that for any sequence $\theta_n = O(n^{-c})$ for some $c > 1/4$, the density function f_ε is bounded and $E||x||^{2+\delta} < \infty$ for some $\delta > 0$. Then*

$$\sup_{\theta_n,t} R_n(\theta_n, t) = \sup_{\theta_n,t} |\frac{1}{\sqrt{n}} \sum_{j=1}^{n} (x_j - Ex)\{I(\varepsilon_j - \theta_n^\tau(x_j - Ex) \le t) - I(\varepsilon_j \le t)$$
$$- F_\varepsilon(t + \theta_n^\tau(x_j - Ex)) + F_\varepsilon(t)\}|$$
$$=: \sup_{\theta_n,t} |\frac{1}{\sqrt{n}} \sum_{j=1}^{n} g_n(x_j, \varepsilon_j, \theta_n, t)| \longrightarrow 0, a.s. \qquad (4.6.1)$$

as $n \to \infty$.

Proof. For any $\eta > 0$, application of Pollard's symmetrization inequality (Pollard, 1984, p. 14) yields for large n

$$P\{\sup_{\theta_n,t} |R_n(\theta_n, t)| \ge \eta\}$$
$$\le 4P\{\sup_{\theta_n,t} |\frac{1}{\sqrt{n}} \sum_{j=1}^{n} \sigma_j g(x_j, \varepsilon_j, \theta_n, t)| \ge \frac{\eta}{4}\} \qquad (4.6.2)$$

provided that for each t and θ_n

$$P\{|R_n(\theta_n, t)| \ge \frac{\eta}{2}\} \le \frac{1}{2}.$$

By the Chebychev inequality and the conditions imposed, the LHS of (4.6.2) is less than or equal to $4\theta_n \text{cov}(x)/\eta$. Hence, (4.6.2) holds for all large n.

To further bound the RHS of (4.6.2) we recall that the class of all functions $g(\cdot,\cdot,\theta_n,t)$ discriminates finitely many points at a polynomial rate; see Gaenssler (1983). An application of the Hoeffding inequality (e.g. see Pollard, 1984, p.16) yields for some $w > 0$

$$P\left\{\sup_{\theta_n,t}\left|\frac{1}{\sqrt{n}}\sum_{j=1}^{n}\sigma_j g(x_j,\varepsilon_j,\theta_n,t)\right| \geq \frac{\eta}{4}\Big|x_1\cdots x_n,\varepsilon_1\cdots\varepsilon_n\right\}$$

$$\leq \left(cn^w \sup_{\theta_n,t}\exp\left[-\frac{\eta^2}{32\sup_t\frac{1}{n}\sum_{j=1}^{n}g^2(x_j,\varepsilon_j,\theta_n,t)}\right]\right)\wedge 1 \quad (4.6.3)$$

where " $\wedge$ " means the minimum. To bound the denominator in the power, similar to Lemma II. 33 in Pollard (1984, p. 31), we derive that for any $c_1 > 0$ there exists a $c_2 > 0$

$$\sup_{\theta_n,t}\frac{1}{n}\sum_{j=1}^{n}|I(\varepsilon_j - \theta_n^\tau(x_j - Ex) \leq t) - I(\varepsilon_j \leq t)|^{c_1} = o_p(n^{-c_2}). \quad (4.6.4)$$

By the Hölder inequality, the sample mean of $g^2(\cdot,\cdot,\theta_n,t)$ is less than or equal to a power of $n^{-1}\sum_{j=1}^{n}(x_j - Ex)^{2+\delta}$ times n^{-c_2} for some $c_2 > 0$. This shows that the RHS of (4.6.3) goes to zero. Integrate out to get the result. Lemma 4.6.1 is proved. □

Proof of Theorem 4.2.1. Slightly modifying the argument of proving Koul's (1992) theorem 2.3.3 or applying Lemma 4.6.1, we can prove the theorem. The details are omitted. □

Proof of Theorem 4.3.1 We deal with I_{n2}^* first. Recall I_{n2}^* has the form, together with (4.3.5),

$$I_{n2}^*(t) = \frac{1}{\sqrt{n}}\sum_{j=1}^{n}x_j^*\{I(\hat{\varepsilon}_j - (\theta_n^*)^\tau x_j^* \leq t), -F_n^*(t)\}$$

where

$$F_n^*(t) = \frac{1}{n}\sum_{j=1}^{n}I(\hat{\varepsilon}_j - (\theta_n^*)^\tau x_j^* \leq t).$$

First of all, we can obtain that for any θ_n with a constraint that $\|\theta_n\| \leq c\log n/n^{\frac{1}{2}}$ and for almost all sequences $\{(x_1,y_1),\cdots,(x_n,y_n),\cdots\}$

$$\begin{aligned}R_n^*(t) &= \frac{1}{\sqrt{n}}\sum_{j=1}^{n}\{[x_j^*(I(\hat{\varepsilon}_j - \theta_n^\tau x_j^* \leq t) - I(\hat{\varepsilon}_j \leq t))\\ &\qquad -E_w[x_j^*(I(\hat{\varepsilon}_j - \theta_n^\tau x_j^* \leq t) - I(\hat{\varepsilon}_j \leq t))]\}\\ &\longrightarrow 0 \qquad \text{a.s.} \end{aligned} \quad (4.6.5)$$

uniformly on t, where E_w stands for the integration over the variable w^*. The argument is very similar to that used to prove Lemma 4.6.1 by noticing that $\theta_n^* = O_p(1/\sqrt{n})$ and letting $\theta_n = \theta_n^*$. Decompose I_{n2}^* as

$$I_{n2}^*(t) = R_n^*(t) + R_{n1}^*(t) + R_{n2}^*(t) - R_{n3}^*(t),$$

where

$$R_{n1}^*(t) = \frac{1}{\sqrt{n}} \sum_{j=1}^n x_j^* \{ I(\hat{\varepsilon}_j \le t) - F_n(t) \},$$

$$R_{n2}^*(t) = \frac{1}{\sqrt{n}} \sum_{j=1}^n \{ E_w x_j^* [I(\hat{\varepsilon}_j - (\theta_n^*)^\tau x_j^* \le t) - I(\hat{\varepsilon}_j \le t)] \}, \quad (4.6.6)$$

$$R_{n3}^*(t) = \frac{1}{\sqrt{n}} \sum_{j=1}^n E_w \{ x_j^* [F_n^*(t) - F_n(t)] \}.$$

Hence we need to show that, combining with (4.6.5), R_{n1}^* converges in distribution to the Gaussian process B, R_{n2}^* converges in distribution to $f_\varepsilon \cdot N$ and R_{n3}^* tends to zero in probability. Invoking Theorem VII 21 of Pollard (1984, p. 157), the convergence of R_{n1}^* can be derived for almost all sequences $\{(x_1, y_1), \cdots, (x_n, y_n), \cdots \}$. The basic steps are as follows: First, we show that the covariance function of R_{n1}^* converges almost surely to that of B. This is easy to do via a little elementary calculation. Second, we check that the conditions in Pollard's Theorem VII. 21 are satisfied, mainly condition (22) on p. 157 (Pollard, 1984). Similar to Lemma 4.6.1 the class of all functions $x^*(I(\hat{\varepsilon} \le \cdot) - F_n(\cdot))$ (this class depends on n) discriminates finitely many point at a polynomial rate, see Gaenssler (1983). The condition (22) is satisfied by applying Lemma VII 15 of Pollard (1984, p. 150) (the definition of covering integral is also on p. 150). Owing to the above description, we omit the details of the proof. The convergence of R_{n3}^* is much easier to obtain as long as we notice that $\sqrt{n}(F_n^* - F_n)$ and $\frac{1}{\sqrt{n}} \sum_{j=1}^n x_j^*$ has a finite limit, the conclusion then holds. The work remaining is to deal with R_{n2}^*. Let

$$R_{n21}^*(t)$$

$$= E_w \left\{ \frac{1}{\sqrt{n}} \sum_{j=1}^n x_j^* [(I(\hat{\varepsilon}_j - (\theta_n^*)^\tau x_j^* \le t) - F_\varepsilon(t + (\beta_n - \beta)^\tau (x_j - \bar{x}) + (\theta_n^*)^\tau x_j^*) \right.$$

$$\left. - (I(\hat{\varepsilon}_j \le t) - F_\varepsilon(t + (\beta_n - \beta)^\tau (x_j - \bar{x})))] \right\}$$

$$=: E_w \left\{ \frac{1}{\sqrt{n}} \sum_{j=1}^n x_j^* [...] \right\}.$$

Noticing $\hat{\varepsilon} = \varepsilon - (\beta_n - \beta)^\tau (x_j - \bar{x})$ and following Lemma II. 33 of Pollard (1984, p. 31), we have that

$$\sup_t \frac{1}{n} \sum_{j=1}^n (x_j^*)^2 [...]^2 = o(n^{-c_2}) \quad a.s.,$$

for almost all sequences $\{(x_1, y_1), \cdots, (x_n, y_n), \cdots\}$ we then easily derive that, similar to Lemma 4.6.1, R_{n21}^* converges in probability to zero uniformly on t. Note that $E_w[x_j^*(I(\hat{\varepsilon}_j \le t) - F_\varepsilon(t + (\beta_n - \beta)(x_j - \bar{x})))] = 0$. Hence

$$R_{n2}^*(t) - R_{n21}^*(t)$$

$$= \frac{1}{\sqrt{n}} \sum_{j=1}^n E_w\Big[x_j^*\{F_\varepsilon(t + (\beta_n - \beta)^\tau(x_j - \bar{x}) + (\theta_n^*)^\tau x_j^*)$$

$$-F_\varepsilon(t + (\beta_n - \beta)^\tau(x_j - \bar{x}))\}\Big]$$

$$= \frac{1}{\sqrt{n}} \sum_{j=1}^n E_w\big[x_j^*(\theta_n^*)^\tau x_j^*\big] f_\varepsilon(t) + o_p(1)$$

$$= E_w\left[\frac{1}{\sqrt{n}} \sum_{j=1}^n (w^*)^2 (x_j - \bar{x})(x_j - \bar{x})^\tau (\theta_n^*)\right] f_\varepsilon(t) + o_p(1)$$

$$=: E_w[...] f_\varepsilon(t) + o_p(1)$$

converges in distribution to $f_\varepsilon \cdot N$ as long as we note the fact that the sum $[...]$ is asymptotically equal to $\frac{1}{\sqrt{n}} \sum_{j=1}^n x_j^* \hat{\varepsilon}_j$ and then is asymptotically normal by the CLT for almost all sequences $\{(x_1, y_1), \cdots, (x_n, y_n), \cdots\}$. The convergence of I_{n2}^* is proved.

We now turn to the proof of the convergence of I_{n3}^*. Note that

$$I_{n3}^*(t) = \frac{1}{\sqrt{n}} \sum_{j=1}^n (x_j - Ex)\{I(\hat{\varepsilon}_j^* - (\beta_n^* - \beta_n)^\tau (x_j - \bar{x}) \le t) - F_n^*(t)\}$$

where

$$F_n^*(t) = \frac{1}{n} \sum_{j=1}^n I(\hat{\varepsilon}_j^* - (\beta_n^* - \beta_n)(x_j - \bar{x}) \le t).$$

Similar to (4.6.6), decompose I_{n3}^* as

$$I_{n2}^*(t) = J_n^*(t) + J_{n1}^*(t) + J_{n2}^*(t) - J_{n3}^*(t)$$

where

$$J_n^*(t) = \frac{1}{\sqrt{n}} \sum_{j=1}^n \{(x_j - Ex)(I(\hat{\varepsilon}_j^* - \theta_n^\tau(x_j - \bar{x}) \le t) - I(\hat{\varepsilon}_j^* \le t)$$

$$-E^*[I(\hat{\varepsilon}_j^* - \theta_n^\tau(x_j - \bar{x}) \le t) - I(\hat{\varepsilon}_j^* \le t)])\},$$

$$J_{n1}^*(t) = \frac{1}{\sqrt{n}} \sum_{j=1}^n (x_j - Ex)\{I(\hat{\varepsilon}_j^* \le t) - F_n(t)\},$$

$$J_{n2}^*(t) = \frac{1}{\sqrt{n}} \sum_{j=1}^{n} (x_j - Ex)E^*[I(\hat{\varepsilon}_j^* - \theta_n^\tau(x_j - \bar{x}) \le t) - I(\hat{\varepsilon}_j^* \le t)]\},$$

$$J_{n3}^*(t) = \frac{1}{\sqrt{n}} \sum_{j=1}^{n} (x_j - Ex)E^*[F_n^*(t) - F_n(t)]\}. \qquad (4.6.7)$$

Similar to (4.6.5) we can derive that, for any θ_n with the constraint that $||\theta_n|| \le c\log n/n^{\frac{1}{2}}$ and for almost all sequences $\{(x_1, y_1), \cdots, (x_n, y_n), \cdots\}$,

$$J_n^*(t) \longrightarrow 0 \qquad \text{a.s.}$$

uniformly on t, where E^* stands for the integration on the bootstrap variable $\hat{\varepsilon}_i^*$. Along with the arguments used for proving I_{n2}^*, we can verify that J_{n1}^* converges in distribution to B, J_{n2}^* converges in distribution to $f_\varepsilon \cdot N$ and J_{n3}^* tends to zero in probability. We omit the details of the proof. □

Proof of Theorem 4.3.2 The argument of the proof is similar to that for Theorem 3.1, hence we only present an outline. Let

$$R_{n4}^* = \frac{1}{\sqrt{n}} \sum_{j=1}^{n} (x_j - \bar{x})\{I(\varepsilon_j^* - (\beta_n^* - \beta_n)^\tau(x_j - \bar{x}) \le t) - I(\varepsilon_j^* \le t)\}.$$

Consider $R_{n4}^* - E_{w^*} R_{n4}^*$ first where

$$E_{w^*} R_{n4}^* = \frac{1}{\sqrt{n}} \sum_{j=1}^{n} (x_j - \bar{x})E_{w_j^*}\{I(w_j^*\hat{\varepsilon}_j - (\beta_n^* - \beta_n)^\tau(x_j - \bar{x}) \le t) - I(w_j^*\hat{\varepsilon}_j \le t)\},$$

and E_{w^*} is the expectation over w^*. Since $\beta_n^* - \beta_n = O(\log n/\sqrt{n})$ a.s., similar to Lemma 4.6.1, for almost all sequences $\{(x_1, y_1) \cdots, (x_n, y_n), \cdots\}$, we can verify that

$$R_{n4}^*(t) - E_{w^*} R_{n4}^*(t) \longrightarrow 0 \text{ a. s.}$$

uniformly on $t \in R^1$. Decompose $I_{n1}^*(t)$ as

$$I_{n1}^*(t) = R_{n4}^*(t) - E_{w^*} R_{n4}^*(t) + R_{n5}^*(t) + E_{w^*} R_{n4}^*(t)$$

where $R_{n5}^*(t) = \frac{1}{\sqrt{n}} \sum_{j=1}^{n} (x_j - \bar{x})I(\varepsilon_j^* \le t)$. We now show that $E_{w^*} R_{n4}^*$ converges to $-f_\varepsilon \cdot N$. Let E_{ε, w^*} denote the expectation on ε and w^*. Define

$$E_{\varepsilon, w^*} R_{n41}^*(t)$$
$$= \frac{1}{\sqrt{n}} \sum_{j=1}^{n} (x_j - \bar{x})[E_{\varepsilon_j, w_j^*} I(w_j^*\hat{\varepsilon}_j - (\beta_n^* - \beta_n)(x_j - \bar{x}) \le t) - E_{\varepsilon_j w_j^*} I(w_j^*\hat{\varepsilon}_j \le t)].$$

Then

$$E_{w^*}R_{n4}^*(t) = \{E_{w^*}R_{n4}^*(t) - E_{\varepsilon,w^*}R_{n41}^*(t)\} + E_{\varepsilon,w^*}R_{n41}^*(t).$$

Along with the argument of proving Lemma 4.6.1 again, we can derive that $\{E_{w^*}R_{n4}^*(t) - E_{\varepsilon,w^*}R_{n41}^*(t)\} \longrightarrow 0$ a.s. uniformly on $t \in R^1$. Now consider $E_{\varepsilon,w^*}R_{n41}^*(t)$. Note that for each j, $E_{w_j^*}I(w_j^*\hat\varepsilon_j \le t) = 1/2I(\hat\varepsilon_j \le t) + 1/2I(-\hat\varepsilon_j \le t)$ and then

$$E_{\varepsilon_j,w_j^*}I(w_j^*\hat\varepsilon_j \le t) = \frac{1}{2}F_\varepsilon\left(\frac{t - (\beta_n - \beta)_{(j)}^\tau(x_j - \bar x)}{1 - (x_j - \bar x)^\tau S_n^{-1}(x_j - \bar x)}\right)$$
$$+ \frac{1}{2}\left(1 - F_\varepsilon\left(\frac{-t + (\beta_n - \beta)_{(j)}^\tau(x_j - \bar x)}{1 - (x_j - \bar x)^\tau S_n^{-1}(x_j - \bar x)}\right)\right)$$

where $(\beta_n - \beta)_{(j)} = S_n^{-1}\sum_{i\ne j}(x_j - \bar x)\varepsilon_i$. Taylor expansion yields that

$$E_{\varepsilon w^*}(R_{n41}^*(t))$$
$$= -\frac{1}{2\sqrt n}\sum_{j=1}^n(x_j - \bar x)(x_j - \bar x)^\tau(\beta_n - \beta)(f_\varepsilon(t) + (f_\varepsilon(-t)) + o_p(1) \text{ a.s.}$$
$$= -f_\varepsilon(t)\cdot N + o_p(1) \text{ a.s.}$$

The last equation is due to the symmetry of f_ε. Now we are in the position to show that R_{n5}^* does not converge in distribution to the Gaussian process B so that the conclusion of the theorem is reached. We can see this immediately by calculating the variance of R_{n5}^* at each t. Actually,

$$\lim_{n\to\infty}\text{Var}(R_{n5}^*(t)) = \frac{1}{4}E(I(\varepsilon \le t) - I(-\varepsilon \le t))^2$$

which is not equal to $\text{Var}(B(t)) = F_\varepsilon(t)(1 - (F_\varepsilon(t))$. The proof of the theorem is completed. $\qquad\square$

5

Checking the Adequacy of a Partially Linear Model

5.1 Introduction

In this chapter, we consider the hypothesis testing for a partially linear model. It is defined by

$$Y = \beta' X + g(T) + \varepsilon$$

where X is d-dimensional random vector, T is d_1-dimensional, β is an unknown parameter vector of d-dimension, $g(\cdot)$ is an unknown measurable function, and the conditional expectation of ε given (T, X) equals zero. Without loss of generality, it is assumed throughout this chapter that X has zero mean. There are many proposals in the literature for the estimation of β and g. Among others, for instance, Cuzick (1992), Engle *et al.* (1986), Mammen and van de Geer (1997), Speckman (1988).

When fitted with independent observations $(t_1, x_1, y_1), \cdots, (t_n, x_n, y_n)$, checking the adequacy of the above fitted model is important and relevant. In this chapter, we consider the null hypothesis as

$$H_0 : E(Y|X = \cdot, T = \cdot) = \alpha + \beta' \cdot + g(\cdot) , \quad \text{for some } \alpha, \beta \text{ and } g \quad (5.1.1)$$

against the saturated alternative

$$H_1 : E(Y|X = \cdot, T = \cdot) \neq \alpha + \beta' \cdot + g(\cdot) , \quad \text{for any } \alpha, \beta \text{ and } g$$

For parametric models, many proposals have been recommended in the literature. For instance, among others, Dette (1999) suggested a test based on the difference of variance estimators, Eubank and Hart (1992) studied the test of score type, Eubank and LaRiccia (1993) proposed a method through variable selection, Härdle and Mammen (1993) considered a test statistic of difference between the parametric and nonparametric fit, Stute (1997) investigated a nonparametric principal component decomposition and derived some optimal tests when the covariate is one-dimensional, Stute, González Manteiga and Presedo Quindimil (1998) applied the test based on residual-marked process, Stute, Thies and Zhu (1998) and Stute and Zhu (2002) proposed an innovation approach to determine p-values conveniently and Fan and Huang (2001) suggested an adaptive Neyman test and Stute and Zhu (2005) recommended a score type test for single-index models. Hart (1997) contains fairly comprehensive references. Most of materials come from Zhu and Ng (2003).

Our test is based on a residual marked empirical process. In the literature, there are several approaches available for constructing test statistics. The main reasons that we use the residual-marked process approach are as follows: since the setting we study is with multivariate covariate, the locally smoothing test constructed by the difference of parametric and nonparametric fits (Härdle and Mammen (1993)) may suffer dimensionality problems because the local smoothing for the nonparametric fit is involved. As for the adaptive Nyeman test (Fan and Huang (2001)), we have not yet known whether it is an asymptotic distribution free test because a nonparametric estimate of $g(\cdot)$ would make the asymptotic null distribution intractable. Moreover, its power performance highly depends on the smoothness of $\varepsilon_j = y_j - \beta' x_j - g(t_j)$ as a function of j. As Fan and Huang (2001) pointed out, making such a smoothness with multivariate predictors is very challenging.

On the other hand, although the globally smoothing tests based on residual marked empirical process have the drawback that the tests are less sensitive to the alternatives of regression functions with the form of oscillation, namely high-frequency, it still shares some desired features as

- the test is consistent for all global alternatives;
- the test is able to detect local alternatives of the order arbitrarily close to $n^{-1/2}$;
- the test is the asymptotically distribution-free (so that the information on the error distribution is not required);
- only lower-dimensional nonparametric function estimate is required for the computation.

As known, the rate of $n^{-1/2}$ is the possibly fastest achievable rate for lack of fit tests in the literature. The adaptive optimal rate of the adaptive Neyman test is $O(n^{-2s/(4s+2)}(\log \log n)^{s/(4s+1)})$ for some $s > 0$ (see Spokoiny (1996) or Fan and Huang (2001)). This means that, theoretically, the test proposed is more sensitive to some local alternative. The fourth feature is desired for multivariate regression problem.

Furthermore, since the exact or limiting null distribution of the test statistic is intractable for computing p-values, we also have to use Monte Carlo approximation, like we do in Chapter 4. In the circumstance we study herein, the situation is complicated. The classical bootstrap approximation has been proved to be inconsistent in the parametric case (see Stute, González Manteiga and Presedo Quindimil (1998)). Even for the wild bootstrap approximation, we are not sure whether it is consistent or not because in related works of testing for the adequacy of parametric models in Chapter 4 and heteroscedasticity checking in Chapter 6, it is shown not to work.

In order to tackle this problem, we apply the NMCT proposed in Section 1.2.3 of Chapter 1.

5.2 A Test Statistic and Its Limiting Behavior

5.2.1 Motivation and Construction

For any weight function $w(\cdot)$, let

$$U(T, X) = (X - E(X|T)), \quad V(T, Y) = (Y - E(Y|T)), \qquad (5.2.1)$$

and

$$\beta = S^{-1}E[U(T, X)V(T, Y)w^2(T)], \quad \gamma(t) = E(Y|T = t), \qquad (5.2.2)$$

with a positive definite matrix $S = E(UU'w^2(T))$. Then it is clear that H_0 is true if and only if

$$E(Y|X, T) = \beta'U(T, X) + \gamma(T).$$

That is, H_0 holds if and only if

$$E[(Y - \beta'U(T, X) - \gamma(T))|(X, T)] = 0.$$

This implies that, for all t, x,

$$E[Y - \beta'U(T, X) - \gamma(T)]w(T)I(T \leq t, X \leq x) = 0, \qquad (5.2.3)$$

where "$X \leq x$" means that each component of X is less than or equal to the corresponding component of x and similarly for "$T \leq t$". The empirical version of LHS of (5.2.3) based on these observations is defined as

$$\frac{1}{n}\sum_{j=1}^{n}[y_j - \beta'U(t_j, x_j) - \gamma(t_j)]w(t_j)I(t_j \leq t, x_j \leq x)$$

which should be close to zero under H_0. When all unknowns are replaced by consistent estimators, we can have the residuals $\hat{\varepsilon}_j = y_j - \hat{\beta}'\hat{U}(t_j, x_j) - \hat{\gamma}(t_j)$. The estimation of the unknown parameter β and functions $\gamma(\cdot)$ and $E(X|T = \cdot)$ will be studied in the next section. We consider a residual marked empirical process

$$R_n(t, x) = \frac{1}{\sqrt{n}}\sum_{j=1}^{n}\hat{\varepsilon}_j w(t_j)I(t_j \leq t, x_j \leq x). \qquad (5.2.4)$$

The proposed test statistic is defined as

$$CV_n = \int (R_n(T, X))^2 d\,F_n(T, X) \qquad (5.2.5)$$

where F_n is the empirical distribution based on $\{(t_1, x_1), \cdots, (t_n, x_n)\}$. We should reject the null hypothesis for the large values of CV_n.

It is worthwhile to mention that CV_n is not a scale-invariant statistic. Usually a normalizing constant is needed, say the estimator of the limiting variance. When the limiting null distribution of the test, if available, is used for p-values, say Fan and Huang (2001), the selection of a good estimator of variance becomes important under the consideration of power performance. It is however not easy to choose. But in our approximation, we do not need such a normalizing constant because in the NMCT approximation, it keeps constant for given (t_i, x_i, y_i)'s. Therefore it does not have any impact on the conditional distribution of the NMCT test statistic. The detail is presented in Section 3. This should be another merit of the NMCT approximation.

5.2.2 Estimation of β and γ

From (5.2.1) and (5.2.2), we can construct estimators. Define for $i = 1, \cdots, n,$

$$\hat{f}_i(t_i) = \frac{1}{n} \sum_{j \neq i}^{n} k_h(t_i - t_j),$$

$$\hat{E}_i(X|T = t_i) = \frac{1}{n} \sum_{j \neq i}^{n} x_j k_h(t_i - t_j) / \hat{f}_i(t_i),$$

$$\hat{E}_i(Y|T = t_i) = \frac{1}{n} \sum_{j \neq i}^{n} y_j k_h(t_i - t_j) / \hat{f}_i(t_i),$$

$$\hat{U}(t_i, x_i) = x_i - \hat{E}_i(X|T = t_i), \quad \hat{V}(t_i, y_i) = y_i - \hat{E}_i(Y|T = t_i),$$

$$\hat{S} = \hat{E}(\hat{U}\hat{U}'w^2(T)) = \frac{1}{n} \sum_{j=1}^{n} \hat{U}(t_j, x_j)\hat{U}(t_j, x_j)'w^2(t_j)$$

where $k_h(t) = (1/h)K(t/h)$ and $K(\cdot)$ is a kernel function defined in Assumptions 5.5.1. The resulting estimators are

$$\hat{\beta} = (\hat{S})^{-1}\frac{1}{n} \sum_{j=1}^{n} \hat{U}(t_j, x_j)\hat{V}(t_j, y_j)w^2(t_j), \quad \hat{\gamma}(t_i) = \hat{E}_i(Y|T = t_i). \ (5.2.6)$$

We have the following asymptotic results about $\hat{\beta}$ and $\hat{\gamma}$.

THEOREM 5.2.1 *Under conditions 1–6 listed in Section 5.5.1, we have that*

$$\sqrt{n}(\hat{\beta} - \beta) = S^{-1}\frac{1}{\sqrt{n}} \sum_{j=1}^{n} U(t_j, x_j)\varepsilon_j w^2(t_j) + O_p([\frac{1}{h\sqrt{n}} + h^2\sqrt{n}]^{1/2})$$

$$(5.2.7)$$

converges in distribution to $N(0, S^{-1}E[U(T,X)U(T,X)'w^4(T)\varepsilon^2]S^{-1})$, *where* $N(0,\Lambda)$ *stands for the normal distribution with mean zero and covariance matrix* Λ *and for any subset [a,b] with* $0 < a < b < 1$,

$$\sup_{a \le t \le b} |\hat{\gamma}(t) - \gamma(t)| = O_p(\frac{1}{\sqrt{nh}} + h). \tag{5.2.8}$$

5.2.3 Asymptotic Properties of the Test

We now state the asymptotic properties of R_n and CV_n. Let

$$J(T,X,Y,\beta,U,S,F(X|T),t,x)$$
$$= \varepsilon w(T)\Big\{I(T \le t, X \le x) - E\Big[I(T \le t, X \le x)U(T,X)'w(T)\Big]S^{-1}U(T,X)$$
$$-F(X|T)I(T \le t)\Big\},$$

where $F(x|T)$ *is the conditional distribution of* X *given* T.

THEOREM 5.2.2 *Under conditions 1–6 in Section 5.5.1, we have that under* H_0,

$$R_n(t,x) = \frac{1}{\sqrt{n}}\sum_{j=1}^{n} J(t_j, x_j, y_j, \beta, U, S, F(x_j|t_j), t, x,) + o_p(1)$$

converging in distribution to R *in the Skorokhod space* $D[-\infty, +\infty]^{(d+1)}$, *where* R *is a centered continuous Gaussian process with the covariance function, for any* (t_1, x_1), (t_2, x_2)

$$E(R(t_1, x_1)(R(t_2, x_2))) \tag{5.2.9}$$
$$= E\big(J(T,X,Y,\beta,U,S,F(X|T),t_1,x_2,)J(T,X,Y,\beta,U,S,F(X|T),t_2,x_2)\big).$$

Therefore, CV_n *converges in distribution to* $CV := \int R^2(T,X)\,dF(T,X)$ *with* $F(\cdot,\cdot)$ *being the distribution function of* (T,X).

We now investigate how sensitive the test is to alternatives. Consider a sequence of models indexed by n

$$E(Y|X,T) = \alpha + \beta'X + g(T) + g_1(T,X)/\sqrt{n}. \tag{5.2.10}$$

THEOREM 5.2.3 *In addition to the conditions of Theorem 5.2.1, assume that* $g_1(T,X)$ *has zero mean and satisfies the condition: there exists a neighborhood of the origin,* U, *and a constant* $c > 0$ *such that for any* $u \in U$

$$|E(g_1(T,X)|T = t + u) - E(g_1(T,X)|T = t)| \le c|u|, \quad \text{for all } t \text{ and } x.$$

Then under the alternative of (5.2.10), R_n converges in distribution to $R+g_{1*}$, where

$$g_{1*}(t,x)$$
$$= E\Big\{[g_1(T,X) - E(g_1(T,X)|T)]w(T)I(T \le t, X \le x)\Big\}$$
$$-E\Big\{U(T,X)'(g_1(T,X) - E(g_1(T,X)|T))w^2(T)\Big\}S^{-1}$$
$$\times E\Big\{U(T,X)w(T)I(T \le t, X \le x)\Big\}$$

is a non-random shift function and thus CV_n converges in distribution to $\int (B(T,X) + g_{1*}(T,X))^2 d\,F(T,X)$.

From the expression of g_{1*}, we realize that it cannot be null unless $g_1(T,X) = \beta_0'X$. Hence the test CV_n is capable of detecting the local alternative distinct arbitrarily close to $n^{-1/2}$ from the null. From the proof of the theorem in the Appendix, it is easy to see that the test is consistent against any global alternative such that $g_1(T,X)w(T)$ is not constant function with respect to $T \in [a,b]$ and X.

5.3 The NMCT Approximation

Let

$$J_1(T,X,Y,t,x,\beta) = \varepsilon w(T)I(T \le t, X \le x),$$
$$J_2(T,X,Y,t,x,U,S) = \varepsilon w^2(T)E[U(T,X)'w(T)I(T \le t, X \le x)]S^{-1}U(T,X),$$
$$J_3(T,X,Y,t,x,\beta,F_{X|T}) = \varepsilon w(T)F(X|T)I(T \le t),$$

then

$$J(T,X,Y,t,x,\beta,U,S,F_{X|T})$$
$$= J_1(T,X,Y,t,x) - J_2(T,X,Y,U,S,t,x) - J_3(T,X,Y,\beta,F(X|T),t,x).$$

From Theorem 5.2.2, we have that asymptotically

$$R_n(t,x) = \frac{1}{\sqrt{n}} \sum_{j=1}^{n} J(t_j,x_j,y_j,\beta,U,S,F_{x_j|t_j},t,x),$$

Therefore, we can use the NMCT approximation in Subsection 1.2.3. That is, we consider the conditional counterpart of R_n and the procedure of determining p-values as follows:

- **Step 1**. Generate random variables $e_i, i = 1, ..., n$ independent with mean zero and variance one. Let $E_n := (e_1, \cdots, e_n)$ and define the conditional counterpart of R_n as

$$R_n(E_n, t, x) = \frac{1}{\sqrt{n}} \sum_{j=1}^{n} e_j J(t_j, x_j, y_j, \hat{\beta}, \hat{U}, \hat{S}, \hat{F}_{x_j|t_j}, t, x,). \quad (5.3.1)$$

where $\hat{\beta}$, $\hat{U}$, $\hat{S}$, $\hat{F}$ are consistent estimators of the unknowns in R_n. The resultant conditional test statistic is

$$CV_n(E_n) = \int (R_n(E_n))^2 F_n(t, x). \quad (5.3.2)$$

- **Step 2**. Generate m sets of E_n, say $E_n^{(i)}, i = 1, ..., m$ and then to get m values of $CV_n(E_n)$, say $CV_n(E_n^{(i)}), i = 1, ..., m$.
- **Step 3**. The p-value is estimated by $\hat{p} = k/(m+1)$ where k is the number of $CV_n(E_n^{(i)})$'s which are larger than or equal to CV_n. Reject H_0 when $\hat{p} \leq \alpha$ for a designed level α.

The following result states the consistence of the approximation.

THEOREM 5.3.1 *Under either H_0 or H_1 and the conditions in Theorem 5.2.2, we have that for almost all sequences $\{(t_1, x_1, y_1), \cdots, (t_n, x_n, y_n), \cdots\}$, the conditional distribution of $R_n(E_n)$ converges to the limiting null distribution of R_n.*

REMARK 5.3.1 *The conditional distribution of $CV_n(E_n)$ serves for determining p-values of the test; we naturally hope that the conditional distribution can well approximate the null distribution of the test statistic no matter whether the data are under either the null or the alternative. On the other hand, as we do not know the underlying model of the data, when a Monte Carlo approximation is applied, we would take the risk that under the alternative the conditional distribution may be far away from the null distribution of the test. If so, it will make the determination of the p-values inaccurate and deteriorate the power performance. However, Theorem 5.3.1 indicates that the conditional distribution based on the NMCT approximation could get rid of this problem to a certain extent.*

REMARK 5.3.2 *We now give some details to explain why we need not choose a normalizing constant in constructing the test statistic. In view of Theorem 5.2.1, we know that a normalizing constant could be*

$$C_n = \sup_{t,x} \frac{1}{n} \sum_{j=1}^{n} \left(J(t_j, x_j, y_j, \hat{\beta}, \hat{U}, \hat{S}, \hat{F}_{x_j|t_j}, t, x) \right)^2,$$

which is the supremum of the sample variance of $J(T, X, Y, \beta, U, S, F_{X|T}, t, x,)$ over t and x. Looking at (5.3.1), we realize that it keeps constant when the (t_i, x_i, y_i)'s are given. Therefore if this normalizing constant is used in the test statistic, that is, the statistic is CV_n/C_n, then the Monte Carlo value of this will be $CV_n(E_n)/C_n$. For determining p-values, it equivalent to CV_n and its conditional counterpart $CV_n(E_n)$.

5.4 Simulation Study and Example

5.4.1 Simulation Study

In the simulations we conducted, the underlying model was

$$y = \beta x + bx^2 + (t^2 - 1/3) + \sqrt{12}(t - 1/2)\varepsilon \qquad (5.4.1)$$

where t is uniformly distributed on $[0,1]$, x and ε are random variables. We considered four cases: 1). Uni-Uni: both x and ε are uniformly on $[-.05, 0.5]$; 2). Nor-Uni: standard normal x and uniform ε on $[-0.5, 0.5]$; 3). Nor-Nor: standard normal x and standard normal ε; 4). Uni-Nor: uniform x on $[-0.5, 0.5]$ and standard normal ε. The empirical powers of these four cases are plotted in Figure 5.1. In the simulations, we chose $\beta = 1$ and $b = 0.0, 0.5, 1.0, 1.5$ and 2.0 for showing the power performance for different alternatives. Note that $b = 0.0$ corresponds to H_0. The sample size was 100 and the nominal level was 0.05. The experiment was performed 3000 times. We chose $K(t) = (15/16)(1 - t^2)^2 I(t^2 \le 1)$ as the kernel function; it has been used by, for example, Härdle (1990) for estimation and Härdle and Mammen (1993) for hypothesis testing. Bandwidth selection is a concern in hypothesis testing. Fan and Li (1996), a relevant work, did not discuss this issue at all. Gozalo and Linton (2001) employed generalized cross-validation (GCV) to select the bandwidth without arguing its use. Eubank and Hart (1993) stated that with homoscedastic errors GCV is useful, while with heteroscedastic errors its usefulness is not clear. Selecting a bandwidth in hypothesis testing is still an open problem and is beyond the scope of this chapter. In our simulation, in order to obtain some insight on how the bandwidth should be chosen, we combined GCV with a grid search. We first computed the average value of h, h_{egcv}, selected by GCV over 1000 replications, then we performed a grid search over $[h_{egcv}-1, h_{egcv}+1]$. For the cases with uniform error, $h = 0.30$ worked best, with $h = 0.57$ best for the cases with normal error. The size is close to the target value of 0.05. The following table only reports the results with these h's.

Table 5.1. Empirical powers of test CV_n with $\alpha = 0.05$

	b	0.00	0.50	1.00	1.50	2.00
Uni-Uni case	$h = 0.30$	0.0460	0.2350	0.4880	0.7940	0.9600
	h_{egcv}	0.0440	0.2370	0.4970	0.7950	0.9570
Nor-Uni case	$h = 0.30$	0.0560	0.6500	0.9800	1.0000	1.0000
	h_{egcv}	0.0540	0.6200	0.9600	1.000	1.000
Nor-Nor case	$h = 0.57$	0.0450	0.3350	0.4410	0.5100	0.6040
	h_{egcv}	0.0450	0.3550	0.4440	0.5130	0.6060
Uni-Nor case	$h = 0.57$	0.0600	0.0700	0.1000	0.1900	0.2600
	h_{egcv}	0.057	0.0700	0.0800	0.1800	0.2300

We considered a comparison with Fan and Li's (1996) test (FL). Since the FL test involves kernel estimation with all covariates including t, we used a product kernel, each factor of which was $K(t) = (15/16)(1-t^2)^2 I(t^2 \leq 1)$. In our initial simulation for the FL test, we were surprised to find that the test had almost no power. We found that the estimate of variance being used has severe influence on power performance since, under the alternatives, its value gets fairly large. Based on this observation, we used the estimate of variance under H_0, with some constant adjustment so as to maintain the significance level. The results of the power are reported in Figure 5.1. Since we know of no other tests for partial linearity except the FL test, we also included a comparison with the adaptive Neyman test of Fan and Huang (2001), who reported that the test was able to detect nonparametric deviations from a parametric model with Gaussian error. The estimate of the variance is also the one under the null hypothesis. In Figure 5.1, Adj-FL and Adj-FH stand, respectively, for the FL test and Fan and Huang's test with the adjustment of variance estimation. For ease of the comparison, we plot the power functions in Figure 5.1, which is put at the end of this chapter.

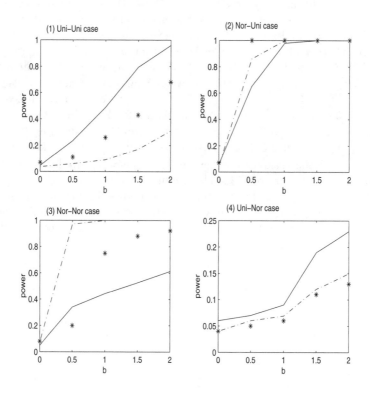

Fig. 5.1. In all plots, the solid line is for the test CV_n with a combined search of GCV and grid points; the dashdot is for the ADJ-FH test; the star is for the ADJ-FL test.

Looking at Figure 5.1(1) – (4), for uniform x, CV_n has higher power than Adj-FL and Adj-FH. The adaptive Neyman test Adj-FH works well with the normal covariate x, see Figure 5.1(2) and (3), while our test does not perform well in this case. After the adjustment, Adj-FL is very sensitive to the alternative in the Nor-Uni case. It seems that there is no uniformly best test here.

5.4.2 An Example

The data are the depths and locations of $n = 43$ earthquakes occurring near the Tonga trench between January 1965 and January 1966 (see Sykes, Isacks and Oliver (1969)). The variable X_1 is the perpendicular distance in hundreds of kilometers from a line that is approximately parallel to the Tonga trench. The variable X_2 is the distance in hundreds of kilometers from an arbitrary line perpendicular to the Tonga trench. The response variable Y is the depth of the earthquake in hundreds of kilometers. Under the plate model, the depths of the earthquakes will increase with distance from the trench and the scatter plot of the data in Figure 5.2 shows this to be the case. Our purpose is to check whether the plate model is linear or not. Looking at Figure 5.2 we find that the plots of Y against X_1 indicate an apparent linear relation with heteroscedasticity, while that between Y and X_2 is not very clear. We find that a linear model is tenable for Y against X_1 with the fitted value $\hat{Y} = -0.295 + 0.949X_1$, while the linear model for Y against X_2 is rejected. If we try a linear model for Y against both X_1 and X_2, $Y = \hat{\beta}_0 + \hat{\beta}^\tau X$ where $X = (X_1, X_2)^\tau$, Figure 5.2 (5) and (6) would support the linearity between Y and X, and the residual plot against $\hat{\beta}^\tau X$ shows a similarity to that of Y against X_1 in Figure 5.2 (2). This finding may be explained as a greater impact on Y by X_1 than by X_2. However in Figure 5.2 (4), we can see that there is some curved structure between the residuals and X_2. The effect of X_2 is not negligible. These observations lead to a more complicated modeling.

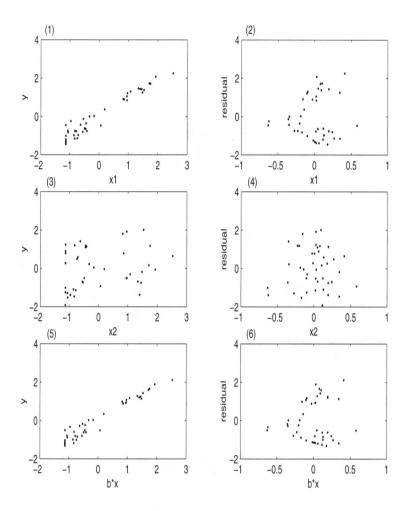

Fig. 5.2. (1), (3) and (5) are scatter plots of Y against X_1, X_2 and $\hat{\beta}^\tau X$, where $\hat{\beta}$ is the least squares estimator of β; (2), (4) and (6) are the residual plots against X_1, X_2 and $\hat{\beta}^\tau X$ when a linear model is used.

A partially linear model $Y = \beta_0 + \beta_1 X_1 + g(X_2) + \varepsilon = \beta_0 + \beta_1 U + r(X_2) + \varepsilon$, with $U = X_1 - E(X_1|X_2)$, provides some reasonable interpretation. Looking at Figure 5.3 (b), we would have $E(X_1|X_2)$ a nonlinear function of X_2. Checking Figure 5.3 (c), $(Y - \beta_0 - \beta U)$ should be nonlinear. The residual plot against

X_2 in this modeling shows that there is no clear indication of relation between the residual and X_2. Using the test suggested in the present chapter, we have $T_n = 0.009$ and the p-value is 0.90. A partial linear model is tenable.

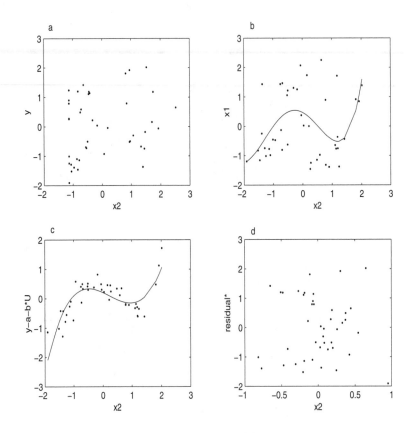

Fig. 5.3. (a). Scatter plot of Y against X_2; (b). Scatter plot of X_1 against X_2 and the fitted curve $\hat{E}(X_1|X_2)$ (the solid line); (c). Scatter plot of $Y - \beta_0 - \beta U$ where $U = X_1 - E(X_1|X_2)$ and the fitted curve of $E(Y|X_2)$ (the solid line); (d). The residual plot against X_2 when the data are fitted by partially linear model $Y = \beta_0 + \beta X_1 + g(X_2)$.

5.5 Appendix

5.5.1 Assumptions

The following conditions are required for the above theorems.
1) Write the first derivative of $E(Y|T = t)$ as $E^{(1)}(Y|T = t)$. Assume that $E^{(1)}(Y|T = t)$, $E(X|T = t)$ and the conditional distribution function of X

given $T = t$, $F(x|t)$ say, all satisfy the following condition: there exists a neighborhood of the origin, say U, and a constant $c > 0$ such that for any $u \in U$ and all t and x

$$|E(X|T = t + u) - E(X|T = t)| \leq c|u|;$$
$$|E^{(1)}(Y|T = t + u) - E^{(1)}(Y|T = t)| \leq c|u|;$$
$$|F(x|t + u) - F(x|t)| \leq c|u|. \qquad (5.5.1)$$

2) $E|Y|^4 < \infty$ and $E|X|^4 < \infty$.
3) The continuous kernel function $K(\cdot)$ satisfies the following properties:
  a) the support of $K(\cdot)$ is the interval $[-1, 1]$;
  b) $K(\cdot)$ is symmetric about 0;
  c) $\int_{-1}^{1} K(u)du = 1$, and $\int_{-1}^{1} |u|K(u)du \neq 0$.
4) As $n \to \infty$ $\sqrt{n}h^2 \to 0$ and $\sqrt{n}h \to \infty$
5) $E(\varepsilon^2|T = t, X = x) \leq c_1$ for some c_1 and all t and x.
6) The weight function $w(.)$ is bounded and continuous on its support set [a,b] with $-\infty < a < b < \infty$ on which the density function $f(\cdot)$ is bounded away from zero.

REMARK 5.5.1 *For the conditions imposed, we can see conditions 1) and 3) are typical for convergence rate of the involved nonparametric estimates. Condition 2) is necessary for asymptotic normality of a least squares estimate. Condition 4) ensures the convergence of the test statistic. Condition 6) is to avoid the boundary effect when a nonparametric smoothing is applied.*

5.5.2 Proof for Results in Section 5.2

Proof of Theorem 5.2.1. Note that the convergence of $\hat{E}_i(Y|t)$ and $\hat{E}_i(X|t)$ has been already proved in the literature. With little modification of Stone's (1982) approach or by the argument of Zhu and Fang (1996), we can derive that

$$\max_{a \leq t_i \leq b} |\hat{E}_i(Y|T = t_i) - E(Y|T = t_i)| = O_p(1/\sqrt{n}h + h),$$

$$\max_{a \leq t_i \leq b} |\hat{E}_i(X|T = t_i) - E(X|T = t_i)| = O_p(1/\sqrt{n}h + h), \qquad (5.5.2)$$

where "$|\cdot|$" stands for the Euclidean norm. As for the convergence of $\hat{\beta}$, note that $\hat{V}(t_j, y_j) = y_j - \hat{E}_j(Y|T = t_j) = \beta'\hat{U}(t_j, x_j) + (y_j - \beta'x_j) - (\hat{E}_j(Y - \beta'X)|T = t_j)$. It can be derived that

$$\hat{\beta} = \beta + \hat{S}^{-1}\frac{1}{n}\sum_{j=1}^{n} \hat{U}(t_j, x_j)((y_j - \beta'x_j) - (\hat{E}_j(Y - \beta'X)|T = t_j))w^2(t_j).$$

It is easy to see that, combining with (5.5.2), $\hat{S}$ converges in probability to S. The work remaining is to show that under the conditions, we have

$$\frac{1}{\sqrt{n}} \sum_{j=1}^{n} \hat{U}(t_j, x_j)((y_j - \beta' x_j) - (\hat{E}_j(Y - \beta' X)|T = t_j))w^2(t_j)$$

$$= \frac{1}{\sqrt{n}} \sum_{j=1}^{n} U(t_j, x_j)w^2(t_j)\varepsilon_j + O_p\left(\left(\frac{1}{h\sqrt{n}} + h^2\sqrt{n}\right)^{1/2}\right). \tag{5.5.3}$$

To this end, we need to show that the following three terms are negligible, i.e.,

$$r_1 = \left| \frac{1}{\sqrt{n}} \sum_{j=1}^{n} \left(\hat{U}(t_j, x_j) - U(t_j, x_j)\right) \times \right.$$

$$\left. \left(\hat{E}_j((Y - \beta' X)|T = t_j) - E((Y - \beta' X)|T = t_j)\right)w^2(t_j) \right|$$

$$= O_p(1/(\sqrt{n}h) + \sqrt{n}h^2)^{1/2},$$

$$r_2 = \left| \frac{1}{\sqrt{n}} \sum_{j=1}^{n} U(t_j, x_j)(\hat{E}_j((Y - \beta' X)|T = t_j) - E((Y - \beta' X)|T = t_j))w^2(t_j) \right|$$

$$= O_p(1/(\sqrt{n}h) + h),$$

and

$$r_3 = \left| \frac{1}{\sqrt{n}} \sum_{j=1}^{n} (\hat{U}(t_j, X_j) - U(t_j, X_j))((y_j - \beta' x_j) - E((Y - \beta' X)|T = t_j))w^2(t_j) \right|$$

$$= O_p(1/(\sqrt{n}h) + h).$$

Without loss of generality, assume that X is scalar in the following. Since the data are independent, $E(\varepsilon|X, T) = 0$ and $(Y - \beta' X) - E((Y - \beta' X)|T) = \varepsilon$, it is easy to see that for $i \neq j$,

$$\left| E\left(((\hat{U}(t_j, x_j) - U(t_j, x_j))w^2(t_j)\varepsilon_i)((\hat{U}(t_i, x_i) - U(t_i, x_i))w^2(t_j)\varepsilon_j)\right) \right| = 0.$$

Together with conditions 5) and 6),

$$E(r_3^2) \leq c_1 E(\hat{U}(T_1, X_1) - U(T_1, X_1))^2 = c_1 E\{(\hat{E}_1(X_1|T_1) - E(X_1|T_1))^2\}$$
$$= O(1/(nh) + h^2).$$

Hence r_3 converges in probability to zero at the rate $O_p(\frac{1}{\sqrt{n}h} + h)$. Let $\hat{g}_i := \hat{E}_i[(Y - \beta' X)|T = t_i]$. Note that $Y - \beta' X = g(T) + \varepsilon$ is only dependent on T and ε and the conditional expectation of $U(T_i, X_i)$ given T_j's and X_j, $j \neq i$ equals zero. Combining the zero conditional expectation of $U(T, X)$ given T, it is easy to see that

$$\left| E[(U(T_i, X_i)(\hat{g}_i - g)w^2(t_j))(U(T_j, X_j)(\hat{g}_j - g)w^2(t_j))] \right| = 0 \qquad i \neq j.$$

Similar to the proof for r_3, we have

$$
\begin{aligned}
E(r_2^2) &\leq E[U^2(T,X)(\hat{g}(T) - g(T))^2 w^4(T)] \\
&\leq \sup_t [(\hat{g}(t) - g(t))^2 w^2(t)] E[U^2(T,X) w^2(T)] \\
&= O(1/(nh) + h^2).
\end{aligned}
\tag{5.5.4}
$$

Hence $r_2 = O_p(1/\sqrt{nh} + h)$. We now turn to prove that r_1 goes toward zero as n tends to infinity. This can be done by using the Cauchy inequality:

$$
\begin{aligned}
r_1^2 &\leq \sqrt{n} \sqrt{\frac{1}{n} \sum_{j=1}^{n} (\hat{U}(t_j, x_j) - U(t_j, x_j))^2 w^2(t_j)} \times \sqrt{\frac{1}{n} \sum_{j=1}^{n} \hat{g}_j^2 w^2(t_j)} \\
&\leq O_p(1/(\sqrt{n}h) + \sqrt{n}h^2)
\end{aligned}
$$

since

$$
\begin{aligned}
&\sup_{i,x} |(\hat{U}(t_i, x) - U(t_i, x)) w(t_i)| \\
&= \sup_i |(\hat{E}_i(X|t_i)) - E(X|t_i)) w(t_i)| = O_p(1/(\sqrt{n}h) + h).
\end{aligned}
$$

The proof is then concluded from condition 4). $\qquad\square$

Proof of Theorem 5.2.2. We prove the results under the null and alternative separately. Consider the case with the null hypothesis first. By Theorem 5.2.1 and a little elementary calculation, it can be shown that

$$
\begin{aligned}
R_n(t, x) &= \frac{1}{\sqrt{n}} \sum_{j=1}^{n} \varepsilon_j w(t_j) I(t_j \leq t, x_j \leq x) \\
&\quad - E(U(T,X)' w(T) I(T \leq t, X \leq x)) S^{-1} \frac{1}{\sqrt{n}} \sum_{j=1}^{n} U(t_j, x_j) \varepsilon_j w^2(t_j) \\
&\quad - \frac{1}{\sqrt{n}} \sum_{j=1}^{n} \hat{g}(t_j) w(t_j) I(t_j \leq t, x_j \leq x) \\
&\quad + \frac{1}{\sqrt{n}} \sum_{j=1}^{n} g(t_j) w(t_j) I(t_j \leq t, x_j \leq x) + O_p\left(\left(\frac{1}{h\sqrt{n}} + \sqrt{n}h^2\right)^{1/2}\right) \\
&=: I_1(t, x) - I_2(t, x) - I_3(t, x) + I_4(t, x) + O_p\left(\left(\frac{1}{h\sqrt{n}} + \sqrt{n}h^2\right)^{1/2}\right).
\end{aligned}
$$

The convergence of I_1 and I_2 just follows the standard theory of empirical process, see, e.g. Pollard (1984, Chapter VII). We now prove that $I_3 - I_4$ converges in distribution to a Gaussian process. Deal with I_3. By a little elementary calculation, we can derive that, letting $\hat{r}_i(t_i) = \frac{1}{n} \sum_{j\neq i}^{n} (y_j - \beta' x_j) k_h(t_i - t_j)$, and invoking (5.5.2), for $a \leq t_j \leq b$

$$\hat{g}(t_j) = \frac{\hat{r}_j(t_j)}{\hat{f}_j(t_j)} = \frac{\hat{r}_j(t_j)}{f(t_j)} + \frac{r(t_j)}{f(t_j)} \frac{f(t_j) - \hat{f}_j(t_j)}{f(t_j)}$$

$$+ \frac{r(t_j)}{f(t_j)} \frac{(f(t_j) - \hat{f}_j(t_j))^2}{\hat{f}_j(t_j)f(t_j)} + \frac{\hat{r}_j(t_j) - r(t_j)}{f(t_j)} \frac{f(t_j) - \hat{f}_j(t_j)}{\hat{f}_j(t_j)}$$

$$= \frac{\hat{r}_j(t_j)}{f(t_j)} + g(t_j) \frac{f(t_j) - \hat{f}_j(t_j)}{f(t_j)} + O_p(\frac{1}{hn} + h^2),$$

where $g(t_j) = r(t_j)/f(t_j)$. Hence

$$I_3(t, x) = \frac{1}{\sqrt{n}} \sum_{j=1}^{n} \frac{\hat{r}_j(t_j)}{f(t_j)} w(t_j) I(t_j \le t, x_j \le x)$$

$$- \frac{1}{\sqrt{n}} \sum_{j=1}^{n} g(t_j) \frac{\hat{f}_j(t_j)}{f(t_j)} w(t_j) I(t_j \le t, x_j \le x)$$

$$+ \frac{1}{\sqrt{n}} \sum_{j=1}^{n} g(t_j) w(t_j) I(t_j \le t, x_j \le x) + O_p(\frac{1}{h\sqrt{n}} + \sqrt{n}h^2)$$

$$=: I_{31}(t, x) - I_{32}(t, x) + I_{33}(t, x) + O_p(\frac{1}{h\sqrt{n}} + \sqrt{n}h^2).$$

We now rewrite $I_{31}(t, x)$ as a U-statistic. Let $w_1(t) = w(t)/f(t)$ and

$$U_h(t_i, x_i, y_i; t_j, x_j, y_j; t, x)$$
$$= \left[(y_i - \beta'x_i)w_1(t_j)I(t_j \le t, x_j \le x) + (y_j - \beta'x_j)w_1(t_i)I(t_i \le t, x_i \le x) \right] \times$$
$$k_h(t_i - t_j).$$

Applying the symmetry of $k_h(\cdot)$ we have for fixed h (i.e. for fixed n)

$$hI_{31}(t, x) = \frac{1}{2n^{3/2}} \sum_{j=1}^{n} \sum_{i \ne j}^{n} h\, U_h(t_i, x_i, y_i; t_j, x_j, y_j; t, x).$$

For the sake of convenience, let $\eta_j = (t_j, x_j, y_j)$. We define a degenerate U-statistic by

$$\frac{\sqrt{n}}{2n(n-1)} \sum_{j=1}^{n} \sum_{i \ne j}^{n} I'_{31}(\eta_i, \eta_j, t, x)$$

$$=: \frac{n}{n-1} h\, I_{31}(t, x) - E(h\, I_{31}(t, x))$$

$$- \frac{1}{\sqrt{n}} \sum_{j=1}^{n} \{ E_1[h\, U_h(\eta; \eta_j; t, x)] - E[h\, U_h(T, X, Y; T_1, X_1, Y_1; t, x)] \}$$

where "E" stands for the overall expectation over all variables and "E_1" stands for the conditional expectation of η, whose distribution is the same as that of η_j, given other η_j. Note that $E_1\big(I'_{31}(\eta, \eta_j, t, x)\big) = E_1\big(I'_{31}(\eta_j, \eta, t, x)\big) = 0$ and the class $\mathcal{G}_n$ of functions consisting of $hU_h(\cdot, \cdot; t, x) - E_1[hU_h(\eta, \eta_j; t, x)]$ over all t and x is a VC class of functions. Therefore $\mathcal{G}_n$ is P-degenerate with envelope

$$
\begin{aligned}
G_n&(\eta_1, \eta_2) \\
&= \Big|\big[(y_1 - \beta' x_1)w_1(t_2) + (y_2 - \beta' x_2)w_1(t_1)\big]|k((t_1 - t_2)/h)\Big| \\
&\quad + 2\Big|E\big[(Y_1 - \beta' X_1)w_1(T_2) + (Y_2 - \beta' X_2)w_1(T_1)\big]|k(T_1 - T_2)/h)\Big| \\
&\quad + \Big|E\big[(Y_1 - \beta' X_1)w_1(t_2) + (y_2 - \beta' x_2)w_1(T_1)\big]|k(T_1 - t_2)/h)\Big|
\end{aligned}
$$

By Theorem 6 of Nolan and Pollard (1987) on p. 786, we have

$$
E \sup_x |\sum_{i,\, j} I'_{31}(\eta_i, \eta_j, t, x)| \le cE(\alpha_n + \gamma_n J_n(\theta_n/\gamma_n))/n^{-3/2}
$$

$$
J_n(s) = \int_0^s \log N_2(u, T_n, \mathcal{G}_n, G_n)du,
$$

$$
\gamma_n = (T_n G_n^2)^{1/2}, \qquad \alpha_n = \frac{1}{4} \sup_{g \in \mathcal{G}_n} (T_n g^2)^{1/2},
$$

$$
T_n g^2 := \sum_{i \ne j} g^2(\eta_{2i}, \eta_{2j}) + g^2(\eta_{2i}, \eta_{2j-1}) + g^2(\eta_{2i-1}, \eta_{2j}) + g^2(\eta_{2i-1}, \eta_{2j-1})
$$

for any function g, and $N_2(\cdot, T_n, \mathcal{G}_n, G_n)$ is the covering number of $\mathcal{G}_n$ under L_2 metric with the measure T_n and the envelope G_n. As $\mathcal{G}_n$ is the VC class, following the argument of the Approximation lemma II 2.25 of Pollard (1984, p. 27) the covering number $N_2(u(T_n G_n^2)/n^2, T_n/n^2, \mathcal{G}_n, G_n)$ can be bounded by cu^{-w_1} for some positive c and w_1, both being independent of n and T_n. Further in probability for large n

$$
T_n G_n^2 = O(hn^2 \log^2 n) \quad a.s.
$$

Hence for large n, $T_n G_n^2/n^2$ is smaller than 1 as $h = n^{-c}$ for some $c > 0$ and $N_2(u, T_n/n^2, \mathcal{G}_n, G_n) \le cu^{-w_1}$. Note that

$$
N_2(u, T_n, \mathcal{G}_n, G_n) = N_2(u/n^2, T_n/n^2, \mathcal{G}_n, G_n).
$$

We can then derive that

$$
\begin{aligned}
J_n(\theta_n/\gamma_n) &\le J_n(1/4) \\
&= n^2 \int_0^{1/(4n^2)} \log N_2(u, T_n/n^2, \mathcal{G}_1, G)d\,u \\
&= -cn^2 \int_0^{1/(4n^2)} \log u\, d\,u = c \log n
\end{aligned}
$$

and

$$\gamma_n^2 = T_n G_n^2 = O(hn^2 \log^2 n) \quad a.s.$$

Therefore for large n, $E \sup_{t,x} |\sum_{i,j} I'_{31}(\eta_i, \eta_j, t, x)| \leq c\sqrt{h/n} \log n$. Equivalently

$$I'_{31}(t, x) = \frac{1}{\sqrt{n}} \sum_{j=1}^{n} \{E_1[U_h(\eta; \eta_j; t, x)]\} + O_p(\log n/\sqrt{nh}) \qquad (5.5.5)$$

Consequently, noting $\frac{n}{(n-1)} E(I_{31}(t, x)) = \frac{\sqrt{n}}{2} E[U_h(\eta; \eta_1; t, x)]$, we have

$$I_{31}(t, x) = \frac{1}{\sqrt{n}} \sum_{j=1}^{n} E_1 \left[U_h(\eta; \eta_j; t, x) \right]$$

$$- \frac{\sqrt{n}}{2} E \left[U_h(\eta; \eta_1; t, x) \right] + O_p(\frac{1}{\sqrt{nh}}). \qquad (5.5.6)$$

From the definition of U_h, condition 1) and some calculation, we have

$$E[U_h(\eta; \eta_1; t, x)]$$
$$= 2E \left[g(T_1) w_1(T) I(T \leq t, X \leq x) k_h(T_1 - T) \right]$$
$$= 2E[g(T + hu) f(T + hu) w_1(T) I(T \leq t, X \leq x) K(u)]$$
$$= 2E[g(T) w(T) I(T \leq t, X \leq x)] + O(h^2)$$

$$(5.5.7)$$

and

$$E_1 U_h(\eta; \eta_j; t, x)$$

$$= w_1(t_j) I(t_j \leq t, x_j \leq x) \int g(hu + t_j) f(hu + t_j) K(u) du$$

$$+ (y_j - \beta' x_j) \int F(x|T) f(T) w_1(T) I(T \leq t) K((T - t_j)/h)/h \, dT$$

$$=: a_j^{(1)}(t, x) + a_j^{(2)}(t, x)$$

where $F(X|T)$ is the conditional distribution of X given T.
    Let

$$b_j^{(1)}(t, x) = g(t_j) f(t_j) w_1(t_j) I(t_j \leq t, x_j \leq x),$$
$$b_j^{(2)}(t, x) = (y_j - \beta' x_j) F(x|t_j) f(t_j) w_1(t_j) I(t_j \leq t). \qquad (5.5.8)$$

We have from conditions 1)–4) and 6)

$$\sup_{t,x} E(a_1^{(1)}(t, x) - b_1^{(1)}(t, x))^2 = O(h^2),$$

$$\sup_{t,x} E(a_1^{(2)}(t, x) - b_1^{(2)}(t, x))^2 = O(h^2). \qquad (5.5.9)$$

Note that $w_1(t) = w(t)/f(t)$. Then

$$E(b_j^{(1)}(t,x)) = E(b_j^{(2)}(t,x)) = E[g(t)w(T)I(T \le t, X \le x)],$$
$$E(a_j^{(1)}(t,x) + a_j^{(2)}(t,x)) = E[U_h(T,X,Y;T_1,X_1,Y_1;t,x)]. \quad (5.5.10)$$

Let $c_j(t,x) = (a_j^{(1)}(t,x) + a_j^{(2)}(t,x) - b_j^{(1)}(t,x) - b_j^{(2)}(t,x))$. We now show that uniformly over t and x

$$\frac{1}{\sqrt{n}} \sum_{j=1}^{n} (c_j(t,x) - Ec_j(t,x)) = O_p(h^{1/2} \log n + h^2 \sqrt{n}). \quad (5.5.11)$$

Note that

$$\sup_{t,x} Var(c_j(t,x)) \le \sup_{t,x} E(c_j^2(t,x))$$
$$\le 2 \sup_{t,x} E\left[\left(a_1^{(1)}(t,x) - b_1^{(1)}(t,x)\right)^2 + \left(a_1^{(2)}(t,x) - b_1^{(2)}(t,x)\right)^2\right]$$
$$\le O(h^2).$$

Recall the class of all functions $c_j(t,x) = c(t_j, x_j, y_j, t, x)$ with indices (t,x) discriminates finitely many points at a polynomial rate (that is, the class is a VC class), see Gaenssler (1983). The application of symmetrization approach and Hoeffding inequality, see Pollard (1984, p. 14–16), yield that for any $\delta > 0$ and some $w > 0$

$$P\{\sup_{t,x} \frac{1}{\sqrt{n}} \sum_{j=1}^{n} (c_j(t,x) - Ec_j(t,x)) \ge \delta\}$$

$$\le 4E\{P\{\sup_{t,x} \frac{1}{\sqrt{n}} \sum_{j=1}^{n} \sigma_j(c_j(t,x) - Ec_j(t,x)) \ge \delta/4 | (T_j, X_j, Y_j), j = 1, \cdots, n\}\}$$

$$\le E\{(cn^w \sup_{t,x} exp([-\frac{\delta^2}{32\frac{1}{n} \sum_{j=1}^{n} (c_j(t,x) - Ec_j(t,x))^2}]) \wedge 1\}.$$

In order to prove the above to be asymptotically zero, we now bound the denominator in the power. Applying condition 1) and the uniformly strong law of large numbers, see Pollard (1984, p. 25, Chapter II Th. 24)

$$\sup_{t,x} \frac{1}{n} \sum_{j=1}^{n} (c_j(t,x) - Ec_j(t,x))^2 = O_p(h).$$

Let $\delta = h^{1/2} \log n$. This implies (5.5.11). Furthermore by (5.5.9), $\sqrt{n} E[c_i(t,x)] = O(\sqrt{n} h^2) = o(1)$. Together with (5.5.8) and (5.5.11), we see that

$$I_{31}(t,x) = \frac{1}{\sqrt{n}} \sum_{j=1}^{n} (b_j^{(1)}(t,x) + b_j^{(2)}(t,x) - Eg(T)w(T)I(T \le t, X \le x))$$

$$+O_p(\frac{1}{\sqrt{nh}}+h)$$

$$= I_4(t,x) + \frac{1}{\sqrt{n}}\sum_{j=1}^{n}(b_j^{(2)}(t,x) - Eg(T)w(T)I(T \le t, X \le x))$$

$$+O_p(\frac{1}{\sqrt{nh}}+h). \tag{5.5.12}$$

Consider I_{32}. Following exactly the above argument of U-statistic, with using condition 1) and $w_2(\cdot) = w(\cdot)g(\cdot)$ in lieu of $y_j - \beta'x_j$ and $w(\cdot)$ respectively, we can verify that

$$I_{32}(t,x)$$
$$= \frac{1}{\sqrt{n}}\sum_{j=1}^{n}g(t_j)w(t_j)I(t_j \le t, x_j \le x)$$
$$+\frac{1}{\sqrt{n}}\sum_{j=1}^{n}\Big(g(t_j)w(t_j)F(x|t_j)I(t_j \le t) - E\big[g(T)w(T)I(T \le t, X \le x)\big]\Big)$$
$$+O_p(\frac{1}{\sqrt{nh}}+h)$$
$$= I_{33}(t,x)$$
$$+\frac{1}{\sqrt{n}}\sum_{j=1}^{n}\Big(g(t_j)w(t_j)F(x|t_j)I(t_j \le t) - E\big[g(T)w(T)I(T \le t, X \le x)\big]\Big)$$
$$+O_p(\frac{1}{\sqrt{nh}}+h). \tag{5.5.13}$$

By (5.5.5), (5.5.12) and (5.5.13),

$$I_3(t,x) - I_4(t,x) = \frac{1}{\sqrt{n}}\sum_{j=1}^{n}\varepsilon w(t_j)F(x|t_j)I(t_j \le t) + O_p(\frac{1}{\sqrt{nh}}+h).$$

$$\tag{5.5.14}$$

It clearly converges in distribution to a Gaussian process. The proof of the theorem is concluded from (5.5.5) and (5.5.14).

Hence $I_3 - I_4$ is a centered stochastic process converging in distribution to a Gaussian process. The proof is finished. □

Proof of Theorem 5.2.3. First note that from the proof of Theorem 5.2.1

$$\sqrt{n}(\hat{\beta} - \beta) = S^{-1}\{\frac{1}{\sqrt{n}}\sum_{j=1}^{n}U(t_j, x_j)\varepsilon_j w^2(t_j)\} + C_1 + o_p(1)$$

where $C_1 = S^{-1}E[U(T,X)(g_1(T,X) - E(g_1(T,X)|T)w^2(T)]$. Similar to the first part of the proof for the case with the null hypothesis, it is derived that

$$R_n(t,x)$$

$$= \frac{1}{\sqrt{n}} \sum_{j=1}^{n} \varepsilon_j w(t_j) I(t_j \le t, x_j \le x)$$

$$-E(U(T,X)'w(T)I(T \le t, X \le x))S^{-1}[\frac{1}{\sqrt{n}} \sum_{j=1}^{n} U(t_j, x_j)\varepsilon_j w^2(t_j) + C]$$

$$-\frac{1}{\sqrt{n}} \sum_{j=1}^{n} (\hat{g}(t_j) - g(t_j))w(t_j)I(t_j \le t, x_j \le x) + O_p(\frac{1}{h\sqrt{n}} + \sqrt{n}h^2)$$

$$+\frac{1}{n} \sum_{j=1}^{n} (g_1(t_j, x_j) - E(g_1(T,X)|T = t_j))w(t_j)I(t_j \le t, x_j \le x)$$

$$= \frac{1}{\sqrt{n}} \sum_{j=1}^{n} \varepsilon_j w(t_j)I(t_j \le t, x_j \le x)$$

$$-E(U(T,X)'w(T)I(T \le t, X \le x))S^{-1}[\frac{1}{\sqrt{n}} \sum_{j=1}^{n} U(t_j, x_j)\varepsilon_j w^2(t_j)]$$

$$-\frac{1}{\sqrt{n}} \sum_{j=1}^{n} (\hat{g}(t_j) - g(t_j))w(t_j)I(t_j \le t, x_j \le x)$$

$$+g_{1*}(t,x) + O_p(\frac{1}{h\sqrt{n}} + \sqrt{n}h^2)$$

$$=: J_1(t,x) - J_2(t,x) - J_3(t,x) + g_{1*}(t,x) + O_p\left((\frac{1}{h\sqrt{n}} + \sqrt{n}h^2)^{1/2}\right),$$

$$(5.5.15)$$

where $g_{1*}(t,x)$ is defined in Theorem 5.2.3. Let $\tilde{Y} = \beta'X + g(T) + \varepsilon$, $g_2(t) = E(\tilde{Y} - \beta'X|T = t)$ and its estimator, at point t_i, $\hat{g}_2(t_i) = \hat{E}_i(\tilde{Y} - \beta'X|T = t_i)$, similar to that in (5.2.6). It is clear that $g(t_i) = g_2(t_i) + E(g_1(T,X)|T = t_i)/\sqrt{n}$ and $\hat{g}(t_i) = \hat{g}_2(t_i) + \hat{E}_i(g_1(T,X)|T = t_i)/\sqrt{n}$. Furthermore, $\sup_i \hat{E}_i(g_1(T,X)|T = t_i) - E(g_1(T,X)|T = t_i)| \to 0$ in probability as $n \to \infty$. Together with this fact,

$$J_3(t,x)$$

$$= \frac{1}{\sqrt{n}} \sum_{j=1}^{n} (\hat{g}_2(t_j) - g_2(t_j))w(t_j)I(t_j \le t, x_j \le x)$$

$$+\frac{1}{n} \sum_{j=1}^{n} \left[\hat{E}_j(g_1(T,X)|T = t_j) - E(g_1(T,X)|T = t_j)\right] w(t_j)I(t_j \le t, x_j \le x)$$

$$= \frac{1}{\sqrt{n}} \sum_{j=1}^{n} \left[\hat{g}_2(t_j) - g_2(t_j)\right] w(t_j)I(t_j \le t, x_j \le x) + o_p(1).$$

Therefore J_3 is asymptotically equal to $I_3(t, x) - I_4(t, x)$ and J_1 and J_2 are analogous to I_1 and I_2 in (5.5.5). The proof follows from the argument for proving Theorem 5.2.2. □

5.5.3 Proof for Results in Section 5.3

Proof of Theorem 5.3.1. All we need is to show that, for almost all sequences $\{(t_1, x_1, y_1), \cdots, (t_n, x_n, y_n), \cdots\}$ and by Wald's device, i) the covariance function of $R_n(E_n)$ converges to that of R, ii) finite distributional convergence (*Fidis Convergence*) of $R_n(E_n)$ holds for any finite indices $(t_1, x_1), \cdots, (t_k, x_k)$, and iii) uniform tightness is true. The properties i) and ii) are easily verified with the use of the fact that even under the local alternative, $\hat{\beta}$, $\hat{U}$, $\hat{S}$ and $\hat{F}(X|T)$ are consistent to β, U, S and $F(X|T)$. The details are omitted. We present the details of the proof for iii). We first notice that the functions $J(\cdot, t, x)$ over all indices (t, x) is a VC class of functions.

For given $\{(t_1, x_1, y_1) \cdots, (t_n, x_n, y_n)\}$, define the $L^2(P_n)$ seminorm as $d_n((t, s), (t', s')) = (P_n(J(T, X, Y, t, s) - J(T, X, Y, t', s'))^2)^{1/2}$, where P_n is the empirical measure based on $\{(t_1, x_1, y_1) \cdots, (t_n, x_n, y_n)\}$ and for any function $f(\cdot)$ of (T, X, Y), $P_n f(T, X, Y)$ denotes the average value of n values $f(T_1, X_1, Y_1), \ldots, f(T_n, X_n, Y_n)$. For uniform tightness, all we need to do is to prove is the equicontinuity lemma holds true, see Pollard (1984, p. 150). By Theorem VII 21 of Pollard (1984, p. 157), $R_n(E_n)$ converges in distribution to a Gaussian process R. Namely, for any $\eta > 0$ and $\epsilon > 0$, there exists a $\delta > 0$ such that

$$\limsup_{n \to \infty} P\{\sup_{[\delta]} |R_n(E_n, t, s) - R_n(E_n, t', s')| > \eta | T_n, X_n, Y_n\} < \epsilon,$$

(5.5.16)

where $[\delta] = \{((t, s), (t', s')) : d_n((t, s), (t', s')) \leq \delta\}$ and $(T_n, X_n, Y_n) = \{(t_1, x_1, y_1), \cdots, (t_n, x_n, y_n)\}$.

Since the limiting property with $n \to \infty$ is investigated, n will always be considered large enough below to simplify some arguments of the proof. Let $J'(\cdot, t, s)$ be the function $J(\cdot, t, s)$ with all true values in the lieu of the estimators, $\mathcal{G} = \{J'(\cdot, t, s) : t \in R^1, s \in R^d\}$ and $d((t, s), (t', s')) = [P_n(J'(T, X, Y, t, s) - J'(T, X, Y, t', s'))^2]^{1/2}$. By the convergence of all estimators invloved, we have that $\sup_{(t,s),(t',s')} |d_n((t, s), (t', s')) - d((t, s), (t', s'))| \to 0$ in probability. Hence for large n

$$P\{\sup_{[\delta]} |R_n(E_n, t, s) - R_n(E_n, t', s')| > \eta | T_n, X_n, Y_n\}$$

$$\leq P\{\sup_{<2\delta>} |R_n(E_n, t, s) - R_n(E_n, t', s')| > \eta | T_n, X_n, Y_n\} \quad (5.5.17)$$

where $2\delta >= \{(t, s) : d(t, s) \leq 2\delta\}$.

In order to apply the chaining lemma (e.g. Pollard (1984), p. 144), we need to check that

$$P\{|R_n(E_n, t, s) - R_n(E_n, t', s')| > \eta\, d((t, s), (t', s'))|T_n, X_n, Y_n\}$$
$$< 2\exp(-\eta^2/2) \tag{5.5.18}$$

and

$$J_2(\delta, d, \mathcal{G}) = \int_0^\delta \{2\log\{(N_2(u, d, \mathcal{G}))^2/u\}\}^{1/2} du \tag{5.5.19}$$

is finite for small $\delta > 0$ where the covering number $N_2(u, d, \mathcal{G})$ is the smallest m for which there exist m points $t_1, \ldots, t_m$ such that $\min_{1 \le i \le m} d((t, s), (t^i, s^i)) \le u$ for every (t, s). (5.5.18) can be derived by the Hoeffding inequality and (5.5.19) is implied by the fact that $\mathcal{G}$ is a VC class and $N_2(u, d, \mathcal{G}) \le c\, u^w$ for some constants c and w. Invoking the chaining lemma, there exists a countable dense subset $< 2\delta >^*$ of $< 2\delta >$ such that, combining with $J_2(\delta, d, \mathcal{G}) \le c\, u^{1/2}$ for some $c > 0$,

$$P\{ \sup_{<2\delta>^*} \sqrt{n}|R_n(E_n, t, s) - R_n(E_n, t', s')| > 26cd^{1/2}|T_n, X_n, Y_n\}$$
$$\le 2c\delta. \tag{5.5.20}$$

The countable dense subset $< 2\delta >^*$ can be replaced by $< 2\delta >$ itself because $R_n(E_n, t, s) - R_n(E_n, t', s')$ is a right-continuous function with respect to t and s. Together with (5.5.17), the proof is concluded from choosing δ small enough. $\qquad\square$

6

Model Checking for Multivariate Regression Models

6.1 Introduction

Suppose that a response vector $Y = (y_1, \ldots, y_q)^T$ depends on a vector $X = (x_1, \ldots, x_p)^T$ of covariables, where T denotes transposition. We may then decompose Y into a vector of functions $m(X) = (m_1(X), \cdots, m_q(X))^T$ of X and a noise variable ε, which is orthogonal to X, i.e., for the conditional expectation of ε given X, we have $E(\varepsilon|X) = 0$. When Y is unknown, the optimal predictor of Y given $X = x$ equals $m(x)$. Since in practice the regression function m is unknown, statistical inference about m is of importance. In a purely parametric framework, m is completely specified up to a parameter. Fort example, in linear regression, $m(x) = \beta^T x$, where $\beta = (\beta_1, \cdots, \beta_q)$ is an unknown $p \times q$ matrix which needs to be estimated from the available data. More generally, we can study a nonlinear model with $m(x) = \Phi(\beta, x) = (\phi_1(\beta_1, x), \cdots, \phi_q(\beta_q, x))^T$, where the vector of the link function $\Phi(\cdot)$ may be nonlinear but is specified.

As discussed in Chapters 4 and 5, checking the adequacy of parametric models becomes one of central problems in regression analysis because any statistical analysis within the model, to avoid wrong conclusions, should be accompanied by a check of whether the model is valid or not. When the dimension $q = 1$, this testing problem has been studied in Chapters 4 and 5.

Note that all of works mentioned in Chapters 4 and 5 focus on the cases with one-dimensional responses. In principle, the relevant methodologies could be, with some modifications, employed to deal with multivariate regression models. However, we should pay particular attention to the correlation between the components of the response. Any direct extension of existing methodology cannot construct a powerful test. This is one of our goals in studying this problem.

In this chapter, we study a score type test for goodness-of-fit. The limiting behavior will be investigated. To enhance the power, we study the optimal choice of the weight function involved in the test statistic.

The second focus of this chapter is the determination of p-values. If we use the limit distribution to do this, power performance of the test is a concern when the sample size is small or moderate. There are several proposals of Monte Carlo approximations available in the literature such as the time-honored Bootstrap (Efron, 1979). Note that the existing methodologies are either parametric (see, e.g. parametric bootstrap, Beran and Ducharme (1991)) or fully nonparametric. If the distribution of the covariables or errors are semi-structured, we should consider semiparametric methodology to take the structure into account, rather than to simply regard the distribution fully nonparametric, and to construct an approximation to the underlying null distribution of the test. As those investigated in Chapter 2, many commonly used classes of distributions are of semi-structured framework. Therefore, in this chapter, we will employ NMCT developed in Chapter 2 to obtain the conditional counterpart of the test to determine p-values. Furthermore, when we do not have information on the structure of the distribution, NMCT in Section 1.2.3 of Chapter 1 will be applied. The consistency of the approximations will be proved.

We note that in the score type test, the estimation of the limiting covariance matrix has to be involved. This is because without this estimation, the limit null distribution cannot be distribution-free and the test cannot be scale-invariant. However, such a plug-in estimation deteriorates the power performance of the test as the estimator becomes larger under alternative than that under the null. It is very difficult to select an estimator which is not affected by the alternative. Interestingly, NMCT completely eliminates this problem. The estimation for that matrix is unnecessary. This is helpful for enhancing the power.

This methodology can easily be applied to a classical problem with the multivariate linear model. To investigate which covariable(s) insignificantly affects the response, the likelihood ratio test called Wilks lambda is a standard test contained in textbooks. The p-values can be determined by chi-square distribution, see e.g., Johnson and Wichern (1992). When the underlying distribution of the error is normal, the Wilks lambda has been proved to be very powerful. However, it is not true when normality is violated. In this chapter, NMCT will be constructed. We will theoretically and empirically show how the NMCT works. The limited simulations show that the power performance of the NMCT is better than the Wilks lambda even in the normal case.

Most of the materials in this chapter are from Zhu and Zhu (2005).

6.2 Test Statistics and their Asymptotic Behavior

6.2.1 A Score Type Test

Suppose that $\{(\boldsymbol{x}_1, \boldsymbol{y}_1), \cdots, (\boldsymbol{x}_n, \boldsymbol{y}_n)\}$ is a sample drawn from a population which follows the model as:

$$\boldsymbol{Y} = \boldsymbol{m}(\boldsymbol{X}) + \boldsymbol{\varepsilon}, \tag{6.2.1}$$

where $\boldsymbol{\varepsilon} = (\varepsilon_1, \cdots, \varepsilon_q)^T$ is a q-dimensional error vector independent of $\boldsymbol{X}$. For model checking, we want to test the null hypothesis: for some matrix $\boldsymbol{\beta} = (\beta_1, \cdots, \beta_q)^T$ almost surely

$$H_0: \quad \boldsymbol{m}(\cdot) = \varPhi(\boldsymbol{\beta}, \cdot), \tag{6.2.2}$$

where for each i with $1 \leq i \leq q$, $m_i(\cdot) = \phi_i(\beta_i, \cdot)$, versus, for any $\boldsymbol{\beta}$, in probability

$$H_1: \quad \boldsymbol{m}(\cdot) \neq \varPhi(\boldsymbol{\beta}, \cdot).$$

Let $\boldsymbol{e} = \boldsymbol{Y} - \varPhi(\boldsymbol{\beta}, \boldsymbol{X})$. Clearly, under H_0, $\boldsymbol{e} = \boldsymbol{\varepsilon}$ and then $\boldsymbol{E}(\boldsymbol{e}|\boldsymbol{X}) = 0$. It implies that for any q-dimensional weight function $\boldsymbol{W}(\boldsymbol{\beta}, \cdot)$ of $\boldsymbol{X}$, $\boldsymbol{E}(\boldsymbol{e} \bullet \boldsymbol{W}(\boldsymbol{\beta}, \boldsymbol{X})) = 0$ where the dot product " $\bullet$ " stands for the multiplication componentwise. From which, we can define a score type test through an empirical version of $\boldsymbol{E}(\boldsymbol{e} \bullet \boldsymbol{W}(\boldsymbol{\beta}, \boldsymbol{X}))$. Let

$$\boldsymbol{T}_n = \frac{1}{n} \sum_{j=1}^{n} \hat{\boldsymbol{e}}_j \bullet \boldsymbol{W}(\hat{\boldsymbol{\beta}}, \boldsymbol{x}_j), \tag{6.2.3}$$

where $\hat{\boldsymbol{e}}_j = \boldsymbol{y}_j - \varPhi(\hat{\boldsymbol{\beta}}, \boldsymbol{x}_j)$ and $\hat{\boldsymbol{\beta}}$ is a consistent estimator of $\boldsymbol{\beta}$. The resulting test statistic is a quadratic form $\boldsymbol{TT}_n = \boldsymbol{T}_n^T \widehat{\boldsymbol{\Sigma}}^{-1} \boldsymbol{T}_n$ where "T" stands for transposition and $\widehat{\boldsymbol{\Sigma}}$ is a consistent estimator of the covariance matrix of $\boldsymbol{T}_n$.

Clearly, there are three quantities in the test statistic to be selected: two estimators $\hat{\boldsymbol{\beta}}$ and $\widehat{\boldsymbol{\Sigma}}$ and the weight function $\boldsymbol{W}(\boldsymbol{\beta}, \cdot)$. The two estimators are for the consistency of the test statistic to obtain a tractable limit null distribution of $\boldsymbol{TT}_n$. The selection for the weight function has an important role for enhancing power performance of the test. If it is not properly selected, the power would be very bad. For instance, if the model is linear, when the estimator $\hat{\boldsymbol{\beta}}$ is the least squares estimator, and the weight function is selected as a $\boldsymbol{W}(\boldsymbol{x}) = \boldsymbol{x}$, the test will have no power at all because the residuals $\hat{\boldsymbol{e}}_j$ are orthogonal to $\boldsymbol{x}_j$. We will return to this topic in Section 6.3.

6.2.2 Asymptotics and Power Study

To estimate $\boldsymbol{\beta}$, we adopt the least squares estimation. The estimator is defined as the maximizer $\hat{\boldsymbol{\beta}}$, over all $\boldsymbol{\beta}$, of the following sum:

$$\sum_{j=1}^{n} (\boldsymbol{y}_j - \varPhi(\boldsymbol{\beta}, \boldsymbol{x}_j))^T (\boldsymbol{y}_j - \varPhi(\boldsymbol{\beta}, \boldsymbol{x}_j)).$$

Clearly each column $\hat{\beta}_i$ of $\hat{\boldsymbol{\beta}}$, $i = 1, \cdots, q$, is the maximizer of

$$\sum_{j=1}^{n}(\boldsymbol{y}_j^{(i)} - \phi_i(\beta_i, \boldsymbol{x}_j))^T (\boldsymbol{y}_j^{(i)} - \phi_i(\beta_i, \boldsymbol{x}_j))$$

over all β_i.

Under regularity conditions, the estimators $\hat{\beta}_i$ are the solutions of the following equations:

$$\sum_{j=1}^{n} \phi_i'(\beta_i, \boldsymbol{x}_j)(\boldsymbol{y}_j^{(i)} - \phi_i(\beta_i, \boldsymbol{x}_j))^T = 0 \qquad (6.2.4)$$

where ϕ' is the $p \times 1$ derivative vector of ϕ with respect to β_i provided that ϕ_i are differentiable. As we know, each of $\hat{\beta}_i$ has an asymptotically linear representation. For model (6.2.1), $\boldsymbol{y}_j^{(i)} = m_i(x_j) + \boldsymbol{e}_j^{(i)}$, $j = 1, \cdots, n$. Denote by $\boldsymbol{\eta} = \boldsymbol{\Phi}(\boldsymbol{\beta}, \boldsymbol{X}) - \boldsymbol{m}(\boldsymbol{X})$, and $\boldsymbol{\eta}_j = \boldsymbol{\Phi}(\boldsymbol{\beta}, \boldsymbol{x}_j) - \boldsymbol{m}(\boldsymbol{x}_j)$. Then $(\boldsymbol{\eta}_1, \cdots, \boldsymbol{\eta}_n)$ are *i.i.d.* random variables. The asymptotically linear representations of $\hat{\beta}_i$ are as follows:

$$\hat{\beta}_i - \beta_i = \frac{1}{n}\sum_{j=1}^{n} S_{ni}^{-1}\phi_i'(\beta_i, \boldsymbol{x}_j)\boldsymbol{e}_j^{(i)} + \frac{1}{n}\sum_{j=1}^{n} S_{ni}^{-1}\phi_i'(\beta_i, \boldsymbol{x}_j)\boldsymbol{\eta}_j^{(i)} + o_p(1/\sqrt{n})$$

$$=: B_{ni} + C_{ni} + o_p(1/\sqrt{n}) \qquad (6.2.5)$$

where $S_{ni} = \frac{1}{n}\sum_{j=1}^{n}(\phi_i'(\beta_i, \boldsymbol{x}_j))(\phi_i'(\beta_i, \boldsymbol{x}_j))^T$. A relevant work is Stute, Zhu and Xu (2005).

Note that in probability, $B_{ni} = O(1/\sqrt{n})$ and under fixed alternatives, S_{ni} and C_{ni} are consistent; that is, $S_{ni} \to S_i := \boldsymbol{E}\Big((\phi_i'(\beta_i, \boldsymbol{X}))(\phi_i'(\beta_i, \boldsymbol{X}))^T\Big)$, and $C_{ni} \to C_i := S_i^{-1}\boldsymbol{E}\Big(\phi_i'(\beta_i, \boldsymbol{X})\boldsymbol{\eta}^{(i)}\Big)$. Thus, $\hat{\beta}_i$ converges to $\beta_i + C_i$ if $C_i \neq 0$. Clearly, $C_i \neq 0$ corresponds to the alternative H_1. In the following, we study the asymptotic behavior of the test statistics under both H_0 and H_1.

Theorem 6.2.1 states the asymptotic results of $\boldsymbol{T}_n = (T_{n1}, \cdots, T_{nq})^T$.

THEOREM 6.2.1 *Assume that*

1.) The second derivatives of ϕ_i and the first derivative of $\boldsymbol{W}^{(i)}$ are continuous and can be bounded by a function $M(\cdot)$ with $E(M(\boldsymbol{x}))^2 < \infty$.

2.) The second moments of ϕ_i, $\boldsymbol{W}^{(i)}$ and $\boldsymbol{e}^{(i)}$ are finite.

3.) The asymptotic representation of $\hat{\beta}$ in (6.2.5) holds. We have

$$\sqrt{n}T_{ni} =: \frac{1}{\sqrt{n}}\sum_{j=1}^{n} \boldsymbol{V}_j^{(i)}\boldsymbol{e}_j^{(i)} + \frac{1}{\sqrt{n}}\sum_{j=1}^{n} \boldsymbol{V}_j^{(i)}\boldsymbol{\eta}_j^{(i)} + o_p(1). \qquad (6.2.6)$$

where $\boldsymbol{V}_j^{(i)} = \Big(\boldsymbol{W}^{(i)}(\beta_i, \boldsymbol{x}_j)) - E[(\boldsymbol{W}^{(i)}(\beta_i, \boldsymbol{X}))(\phi_i'(\beta_i, \boldsymbol{X}))^T]S_i^{-1}(\phi_i'(\beta_i, \boldsymbol{x}_j))\Big)$.
Then

- Under H_0, $\sqrt{n}(T_{ni} - T_i) \Longrightarrow N(0, \sigma_{ii})$ where the notation "$\Longrightarrow$" stands for weak convergence and σ_{ii} is the variance of $\boldsymbol{V}_j^{(i)} \boldsymbol{e}_j^{(i)}$. Therefore letting $\boldsymbol{T} = (T_1, \cdots, T_q)^T$, $\sqrt{n}(\boldsymbol{T}_n - \boldsymbol{T}) \Longrightarrow N(0, \boldsymbol{\Sigma})$ with $\boldsymbol{\Sigma} = (\sigma_{lm})_{1 \le l, m \le q}$ and σ_{lm} is the covariance between $\boldsymbol{V}_j^{(l)} \boldsymbol{e}_j^{(l)}$ and $\boldsymbol{V}_j^{(m)} \boldsymbol{e}_j^{(m)}$ for any pair of $1 \le l, m \le q$. This results in that $\boldsymbol{T}\boldsymbol{T}_n$ is asymptotically chi-squared with degree q of freedom.
- Under H_1, when for some i with $1 \le i \le q$, if $\left[\frac{1}{\sqrt{n}} \sum_{j=1}^n \boldsymbol{V}_j^{(i)} \boldsymbol{\eta}_j^{(i)} \right]^2 \to \infty$, then $\boldsymbol{T}\boldsymbol{T}_n \to \infty$ in probability; and if $\left[\frac{1}{\sqrt{n}} \sum_{j=1}^n \boldsymbol{V}_j^{(i)} \boldsymbol{\eta}_j^{(i)} \right] \to S_i$, a constant, then T_{ni} converges in distribution to $T_i + S_i$ where $S_i = E\left[\boldsymbol{V}^{(i)} \boldsymbol{\eta}^{(i)} \right]$. Let $\boldsymbol{T} = (T_1, \cdots, T_q)^T$ and $\boldsymbol{S} = (S_1, \cdots, S_q)^T$. $\boldsymbol{T}\boldsymbol{T}_n$ then converges in distribution to $(\boldsymbol{T} + \boldsymbol{S})^T \boldsymbol{\Sigma}^{-1} (\boldsymbol{T} + \boldsymbol{S})$ that is a non-central chi-squared random variable with the non-centrality $\boldsymbol{S}^T \boldsymbol{\Sigma}^{-1} \boldsymbol{S}$.

This theorem means that the test can detect the alternatives distinct $n^{-1/2}$ from the null hypothesis if the non-centrality $\boldsymbol{S}^T \boldsymbol{\Sigma}^{-1} \boldsymbol{S}$ is not zero. The following subsection discusses the selection of $\boldsymbol{W}$.

6.2.3 The Selection of $\boldsymbol{W}$

Consider the case of $q = 1$ first. The distribution of $(\boldsymbol{T} + \boldsymbol{S})^T \boldsymbol{\Sigma}^{-1} (\boldsymbol{T} + \boldsymbol{S})$ is the non-central chi-squared with the non-centrality $\boldsymbol{S}^T \boldsymbol{\Sigma}^{-1} \boldsymbol{S}$. See Stute and Zhu (2005) and Zhu and Cui (2005). When we consider the one-sided test, its power function is $\Phi(-c_\alpha/2 + \boldsymbol{\Sigma}^{-1/2} \boldsymbol{S}) + \Phi(-c_\alpha/2 - \boldsymbol{\Sigma}^{-1/2} \boldsymbol{S})$ where c_α is the upper $(1 - \alpha)$-quantile of the normal distribution. It is easy to prove that this function is a monotone function of $|\boldsymbol{\Sigma}^{-1/2} \boldsymbol{S}|$. This implies that the power function of the test is a monotone function of the non-centrality $\boldsymbol{S}^T \boldsymbol{\Sigma}^{-1} \boldsymbol{S}$. For the multivariate response case, we have the following lemma.

LEMMA 6.2.1 *Under the conditions of Theorem 6.2.1, the power function relating to the distribution of $(\boldsymbol{T} + \boldsymbol{S})^T \boldsymbol{\Sigma}^{-1} (\boldsymbol{T} + \boldsymbol{S})$ is a monotone function of* $\boldsymbol{S}^T \boldsymbol{\Sigma}^{-1} \boldsymbol{S}$.

From Lemma 6.2.1, we can see that to enhance the power, we should select $\boldsymbol{W}$ to allow $\sum_{i=1}^q v_i^2 = \boldsymbol{S}^T \boldsymbol{\Sigma}^{-1} \boldsymbol{S}$ as large as possible.

LEMMA 6.2.2 *Under the conditions of Theorem 6.2.1, the optimal choice of $\boldsymbol{W}$ satisfies the equation that $\boldsymbol{\Sigma}^{-1/2} \boldsymbol{V} = [\boldsymbol{E}(\boldsymbol{\eta}^2)]^{-1/2} \boldsymbol{\eta}$ where $[\boldsymbol{E}(\boldsymbol{\eta}^2)]$ is a diagonal matrix each element on the diagonal is $\boldsymbol{E}(\boldsymbol{\eta}^{(i)})^2$ and of which where $\boldsymbol{V} = (\boldsymbol{V}^{(1)}, \cdots, \boldsymbol{V}^{(q)})^T$ and $\boldsymbol{V}^{(i)}$ are defined in Theorem 6.2.1.*

Remark 6.1. Lemma 6.2.2 provides a way to search for an optimal weight through resolving an equation. In a special case, the solution has a closed form. When W is orthogonal to ϕ', V is actually equal to W. If the components of W are orthogonal to one another, ε is independent of X and has a common variance, σ^2, of all components $e^{(i)}$, we have that Σ is also a diagonal matrix each element on the diagonal being $\sigma^2 E(W^{(i)})$, $i = 1, \cdots, q$. Hence, $\Sigma^{-1/2}V = \sigma^2[E(W^2)]^{-1/2}W = [E(\eta^2)]^{-1/2}\eta$. This means that W can be selected as η because σ^2 is a constant. This selection fits a direct observation that the weight function proportional to η should be a good candidate. Furthermore, the orthogonality of W to ϕ' is a reasonable request because the departure from the null model can be within the space perpendicular to the space spanned by $\phi'(\beta, X)$ for all X. When $\phi'(\beta, X) = X$, a good weight function should be selected within the space which is orthogonal to the linear space spanned by all X. For a univariate response case, a similar discussion can be seen in Zhu and Cui (2005) and Stute and Zhu (2005). On the other hand, when we do not have much prior information on the alternatives, η is unknown and is not even estimable. Thus, this optimal weight cannot be used. In this situation, the desired test should be an omnibus test. For this, a more practically useful method is to use the plots of residuals or of Y against Φ. We will describe this graphical method in Section 6.4.

6.2.4 Likelihood Ratio Test for Regression Parameters

Let us describe the classical diagnostic issue through likelihood ratio. This can be found in any textbook of multivariate analysis, see, e.g. Johnson and Wichern (1992). Consider the linear model

$$Y = \beta^T X + \varepsilon \qquad (6.2.7)$$

where ε is independent of X. To check whether some component of X has impact for Y, we want to test the hypothesis

$$H_0: \ \beta_{(1)}^T = 0,$$

where $\beta^T = (\beta_{(1)}^T, \beta_{(2)}^T)$, $\beta_{(1)}^T$ is a $q \times l$ matrix and $\beta_{(1)}^T$ is a $q \times (p-l)$ matrix. Let $x^T = ((x^{(1)})^T, (x^{(2)})^T)$ with $(x^{(1)})^T$ being a l-dimensional row vector and $(x^{(1)})^T$ being a $(p-l)$-dimensional row vector. Under H_0, the model becomes

$$Y = \beta_{(2)}^T X^{(2)} + \varepsilon.$$

When a sample $\{(x_1, y_1), \cdots, (x_n, y_n)\}$ is available, by least squares estimation, we can separately obtain the least squares estimators $\hat{\beta}^T$ and $\hat{\beta}_{(2)}^T$ of β^T and $\beta_{(2)}^T$ respectively. Hence two sums of squares and cross-products can be derived as

$$\hat{\Sigma} = \sum_{j=1}^{n}(y_j - \hat{\beta}^T x_j)(y_j - \hat{\beta}^T x_j)^T;$$

$$\hat{\Sigma}_2 = \sum_{j=1}^{n}(y_j - \hat{\beta}_{(2)}^T x_j^{(2)})(y_j - \hat{\beta}_{(2)}^T x_j^{(2)})^T.$$

A modified logarithm of the likelihood ratio test, popularly called the Wilks lambda test, is

$$\Lambda_n = -[n - p - 1 - 1/2(q - p + l + 1)] \ln \left(|\hat{\mathbf{\Sigma}}| / |\hat{\mathbf{\Sigma}}_2| \right). \qquad (6.2.8)$$

Under H_0, this statistic converges to a chi-square distribution with $q(p - l)$ degree of freedom.

6.3 NMCT Procedures

From Theorem 6.2.1 for $\boldsymbol{TT}_n$ and the limit behavior of Λ_n, we can easily determine p-values of the tests in Section 2 through chi-square distributions. However, when the sample size is small, the limit distributions do not work well. Furthermore, for $\boldsymbol{TT}_n$, a deterioration for the power comes from a plug-in estimation for the covariance matrix $\boldsymbol{\Sigma} = Cov(\boldsymbol{V} \bullet \boldsymbol{\varepsilon})$. This is because under the alternative $\boldsymbol{\varepsilon}$ is no longer centered and this covariance matrix will be much larger than that under H_0. In the literature, there are many proposals for approximating the null distributions of $\boldsymbol{TT}_n$ and λ_n. Bootstrap is the most popularly used approximation. Since Efron's (1979) time-honored work, many variants of the bootstrap have been appearing in the literature. Among others, especially for model checking, Wild bootstrap (Wu (1986), Härdle and Mammen (1993), Stute, González Manteiga and Presedo Quindimil (1998)) is a good alternative. However, a bootstrap procedure also requires an estimation for Σ.

6.3.1 The NMCT for $\boldsymbol{TT}_n$

The above algorithm cannot be applied directly to the regression problem. Note that the error $\boldsymbol{\varepsilon}$ is unobservable. In other words, it is impossible to generate reference random variables which have the same distribution as $\boldsymbol{\varepsilon}$. Note that our test is residual-based. To approximate the null distribution of the test, we cannot simply simulate residuals like that described in Chapter 2 because under alternative, the distribution of the Monte Carlo test based on the simulated residuals is not an approximation to the null distribution, but an approximation to the distribution under alternative. We have to study the structure of the test first to see how a NMCT should be constructed.

Recalling the asymptotic representation of T_n in (6.2.6), the first term relates to the errors and the second term is associated with the alternative regression function. This motivates us to construct a conditional approximation based on the first term. For the sake of notational simplicity, we only present an algorithm with elliptically symmetric distribution of the error. Similarly for other classes of distributions described in Chapter 2.

- **Step 1.** Generate independent identically distributed random variables $u_i = N_i/\|N_i\|, i = 1, \ldots, n$ where N_i has normal distribution $N(0, I_q)$. Clearly, u_i is uniformly distributed on the sphere surface. Let $U_n := \{u_i, i = 1, \ldots, n\}$ and define the conditional counterpart of T_n as

$$\tilde{T}_n(E_n) = \frac{1}{\sqrt{n}} \sum_{j=1}^{n} \hat{V}_j \bullet u_j \bullet \|\hat{e}_j\|, \qquad (6.3.1)$$

where $\hat{V}_j = \left\{ W(\hat{\beta}, x_j) - \hat{E}[(W(\hat{\beta}, X))(\phi'(\hat{\beta}, X))^T]\hat{S}^{-1}(\phi'(\hat{\beta}, x_j)) \right\}$. The resulting conditional counterpart of the test statistic $TT'_n = T_n^T T_n$ is

$$TT'_n(U) = \left[\tilde{T}_n(U_n) \right]^T \left[\tilde{T}_n(U_n) \right]. \qquad (6.3.2)$$

- **Step 2.** Generate m sets of U_n, say $U_n^{(i)}, i = 1, \ldots, m$ and get k values of $TT'_n(U_n)$, say $(TT'_n(U_n))^{(i)}, i = 1, \ldots, m$.
- **Step 3.** The p-value is estimated by $\hat{p} = k/(m+1)$ where k is the number of $(TT'_n(U_n))^{(i)}$'s which are larger than or equal to TT'_n. Reject H_0 when $p \le \alpha$ for a designated level α.

Remark 6.2. From the above procedure, we can see that we use a NMCT statistic TT'_n which differs from TT_n. Also TT'_n is not scale-invariant. However, when we use this NMCT, any constant scalar is eliminated in terms of comparing the values of $TT'_n(U_n)$ with TT'_n. Therefore, scale-invariance is not important in our case. The following result states the consistency of the approximation.

THEOREM 6.3.1 *Assume that $1/n \sum_{j=1}^{n}(V_j \bullet \eta_j)(V_j \bullet \eta_j)^T$ converges to zero in probability and the conditions of Theorem 6.2.1 hold. Then we have that, for almost all sequences $\{(x_i, y_i), i = 1, \ldots, n, \cdots, \}$, the conditional distribution of $TT'_n(U_n)$ converges to the limiting null distribution of TT_n. When $(1/n) \sum_{j=1}^{n}(V_j \bullet \eta_j)(V_j \bullet \eta_j)^T$ converges in probability to a constant matrix, $TT'_n(U_n)$ converges in distribution to TT which may have a different distribution from the limiting null distribution of TT_n.*

When the distribution of ε does not have a special structure, in other words, ε is fully nonparametric, we can also construct a NMCT based on the idea described in Section 1.2.3 which is easy to implement.

- **Step 1′.** Generate independent identically distributed random vectors $u_i, i = 1, \ldots, n$ with bounded support and mean 0 and covariance matrix **1**. That is, all components are identical. Let $U_n := \{u_i, i = 1, \ldots, n\}$ and define the conditional counterpart of T_n as

$$\tilde{T}_n(U_n) = \frac{1}{\sqrt{n}} \sum_{j=1}^{n} \hat{V}_j \bullet u_j \bullet \hat{e}_j, \qquad (6.3.3)$$

where $\hat{V}_j = \left\{ W(\hat{\beta}, x_j) - \hat{E}[(W(\hat{\beta}, X))(\Phi'(\hat{\beta}, X))^T]\hat{S}^{-1}(\Phi'(\hat{\beta}, x_j)) \right\}$.
The resulting conditional counterpart of TT'_n is

$$TT'_n(U_n) = [\tilde{T}_n(\varepsilon_n)]^T [\tilde{T}_n(U_n)]. \qquad (6.3.4)$$

- **Step 2'**. Generate m sets of U_n, say $U_n^{(i)}, i = 1, \ldots, m$ and get m values of $TT'_n(U_n)$, say $(TT'_n(U_n))^{(i)}, i = 1, \ldots, m$.
- **Step 3'**. The p-value is estimated by $\hat{p} = k/(m+1)$ where k is the number of $(TT'_n(U_n))^{(i)}$'s which are larger than or equal to TT'_n. Reject H_0 when $p \le \alpha$ for a designated level α.

THEOREM 6.3.2 *Assume that* $(1/n) \sum_{j=1}^{n}(V_j \bullet \eta_j)(V_j \bullet \eta_j)^T$ *converges to zero in probability and the conditions in Theorem 6.2.1 hold. Then the conclusion of Theorem 6.3.1 holds.*

6.3.2 The NMCT for the Wilks Lambda

For the Wilks lambda, we also study its structure first. Let $\mathcal{Y} = (y_1, \cdots, y_n)$, $\mathcal{X} = (x_1, \cdots, x_n)$, $\mathcal{X}_{(2)} = (x_1^{(2)}, \cdots, x_n^{(2)})$, $\mathcal{E} = ((e_1, \cdots, e_n))$ be, respectively, the $q \times n$ response matrix, $p \times n$ and $(p - l) \times n$ covariate matrices and $q \times n$ error matrix. Note that $\hat{\beta} = \left(\mathcal{X}\mathcal{X}^T \right)^{-1} \mathcal{X}\mathcal{Y}^T$ and $\hat{\beta}_{(2)} = \left(\mathcal{X}_{(2)}\mathcal{X}_{(2)}^T \right)^{-1} \mathcal{X}_{(2)}\mathcal{Y}^T$.
It is easy to obtain that

$$\mathcal{Y} - \mathcal{Y}\mathcal{X}^T \left(\mathcal{X}\mathcal{X}^T \right)^{-1} \mathcal{X} = \mathcal{E}\left[I - \mathcal{X}^T \left(\mathcal{X}\mathcal{X}^T \right)^{-1} \mathcal{X} \right],$$

$$\mathcal{Y} - \mathcal{Y}\mathcal{X}_{(2)}^T \left(\mathcal{X}_{(2)}\mathcal{X}_{(2)}^T \right)^{-1} \mathcal{X}_{(2)} = \mathcal{E}\left[I - \mathcal{X}_{(2)}^T \left(\mathcal{X}_{(2)}\mathcal{X}_{(2)}^T \right)^{-1} \mathcal{X}_{(2)} \right],$$

where I is a $n \times n$ identity matrix. From the definition of $\hat{\Sigma}$ and $\hat{\Sigma}_2$ of Subsection 6.2.4, the above implies that

$$\hat{\Sigma} = \left[\mathcal{Y} - \mathcal{Y}\mathcal{X}^T \left(\mathcal{X}\mathcal{X}^T \right)^{-1} \mathcal{X} \right]\left[\mathcal{Y} - \mathcal{Y}\mathcal{X}^T \left(\mathcal{X}\mathcal{X}^T \right)^{-1} \mathcal{X} \right]^T$$

$$= \mathcal{E}\left[I - \mathcal{X}^T \left(\mathcal{X}\mathcal{X}^T \right)^{-1} \mathcal{X} \right]\mathcal{E}^T;$$

$$\hat{\Sigma}_2 = \left[\mathcal{Y} - \mathcal{Y}\mathcal{X}_{(2)}^T \left(\mathcal{X}_{(2)}\mathcal{X}_{(2)}^T \right)^{-1} \mathcal{X}_{(2)} \right]\left[\mathcal{Y} - \mathcal{Y}\mathcal{X}_{(2)}^T \left(\mathcal{X}_{(2)}\mathcal{X}_{(2)}^T \right)^{-1} \mathcal{X}_{(2)} \right]^T$$

$$= \mathcal{E}\left[I - \mathcal{X}_{(2)}^T \left(\mathcal{X}_{(2)}\mathcal{X}_{(2)}^T \right)^{-1} \mathcal{X}_{(2)} \right]\mathcal{E}^T.$$

From these two formulae, we now define a NMCT. Like that in Subsection 6.3.1, generate $q \times n$ random matrix $\mathcal{U}_n = (\boldsymbol{u}_1, \cdots, \boldsymbol{u}_n)$; define

$$\hat{\boldsymbol{\Sigma}}(\mathcal{U}_n) = \left(\mathcal{U}_n \bullet \hat{\mathcal{E}}\right) \left[\boldsymbol{I} - \mathcal{X}^T \left(\mathcal{X}\mathcal{X}^T\right)^{-1} \mathcal{X}\right] \left(\mathcal{U}_n \bullet \hat{\mathcal{E}}\right)^T;$$

$$\hat{\boldsymbol{\Sigma}}_2(\mathcal{U}_n) = \left(\mathcal{U}_n \bullet \hat{\mathcal{E}}\right) \left[\boldsymbol{I} - \mathcal{X}_{(2)}^T \left(\mathcal{X}_{(2)}\mathcal{X}_{(2)}^T\right)^{-1} \mathcal{X}_{(2)}\right] \left(\mathcal{U}_n \bullet \hat{\mathcal{E}}\right)^T$$

where " $\bullet$ " stands for dot product, see Section 1.2.2 of Chapter 1. Repeat this step m times to generate m values of $\Lambda_n(\mathcal{U}_n) = -[n - p - 1 - 1/2(q - p + l + 1)] \ln \left(|\hat{\boldsymbol{\Sigma}}(\mathcal{U}_n)|/|\hat{\boldsymbol{\Sigma}}_2(\mathcal{U}_n)|\right)$, say $\Lambda_n(\mathcal{U}_n^{(1)}), \cdots, \Lambda_n(\mathcal{U}_n^{(m)})$; and count the number k of $\Lambda_n(\mathcal{U}_n^{(i)})$'s which are greater than or equal to Λ_n to obtain the estimated p-value $k/(m+1)$.

Similar to Theorem 6.3.1, we have the asymptotic equivalence between $\Lambda_n(\mathcal{U}_n)$ and Λ_n.

THEOREM 6.3.3 *Assume that the fourth moment of $\boldsymbol{X}$ and $\boldsymbol{Y}$ exists. Then for almost all sequences $\{(\boldsymbol{x}_1, \boldsymbol{y}_1), \cdots, (\boldsymbol{x}_n, \boldsymbol{y}_n)\}$, the conditional distribution $\Lambda_n(\mathcal{U}_n))$ converges to the limit distribution of Λ_n.*

6.4 Simulations and Application

6.4.1 Model Checking with the Score Type Test

Example 1. The model is with continuous response, namely

$$\boldsymbol{Y} = (\boldsymbol{\beta}^T \boldsymbol{X}) + c\boldsymbol{X}^2 + \varepsilon \tag{6.4.1}$$

where $\boldsymbol{Y}$ is q-dimensional and $\boldsymbol{X}$ p-dimensional, $\boldsymbol{X}$ and ε are independent, and $\boldsymbol{X}$ is multivariate normal $N(0, \boldsymbol{I}_p)$. To check the performance of the NMCT procedure, we considered three distributions of the error ε: $N(0, I_q)$, normal; $U_q(-0.5, 0.5)$, uniform on the cube $(-0.5, 0.5)^q$, and $\chi_q^2(1)$ all components following chi-square with degree 1 of freedom respectively. The hypothetical model was $\Phi(\boldsymbol{X}) = \boldsymbol{\beta}^T \boldsymbol{X}$ and $s(\boldsymbol{X}) = \boldsymbol{x}^2$. Therefore the null model holds if and only if $c = 0$. In the simulation, we considered $c = 0, 0.1, 0.2, \cdots, 1$ and $p = 3$ and $q = 2$ and the matrix $\boldsymbol{\beta} = B = [1, 0; 1, 1; 0, 1]$.

When we regard the alternatives as directional ones, the weight function can be selected as $\boldsymbol{X}^2$. As we discussed above, the residual plots of ε against $\boldsymbol{X}$ are also informative. We plotted all components of $\boldsymbol{\varepsilon}$ against all components of $\boldsymbol{X}$, and associated linear combinations $\beta_i^T \boldsymbol{X}$. In Figure 6.1, we only report the plots of the residuals against $\beta_i^T \boldsymbol{X}$ with 300 generated data points. The plots indicate a pattern of quadratic curve and then weight function $\boldsymbol{X}^2$ is also suggested. Hence, for this model, we used $\boldsymbol{W}(\boldsymbol{X}) = \boldsymbol{X}^2$.

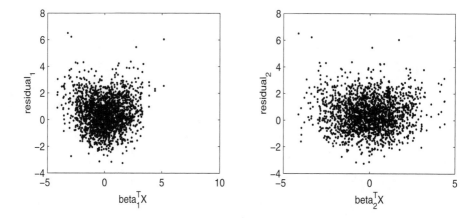

Fig. 6.1. This figure reports the plots of residuals $Y_i - \beta_i^T X$ $i = 1, 2$ against the fitted linear models $\beta_i^T X$ with model (6.4.1) when $c = 0.5$.

Since this is the first research work with multivariate responses in this area, there are no other competitors in the literature; we compared the power performance of the test when the critical values were determined by the limit distribution and the NMCT. The sample sizes were $n = 20, 40, 60$. From Figure 6.2, we can clearly see the superiority of the NMCT when the sample size is $n = 20$. Another interesting observation is that in all cases with different error distributions, the NMCT performs better than the limit distribution although when the sample size is large, the powers with these two testing procedures are very close one another.

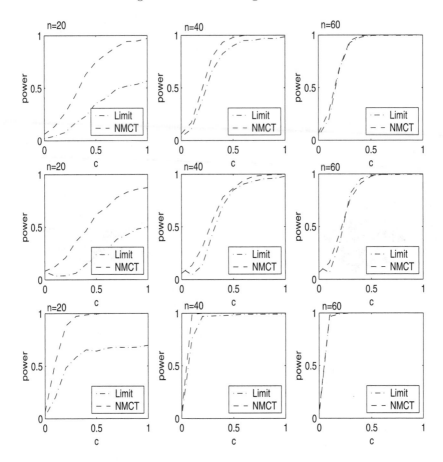

Fig. 6.2. This figure reports the plots of the power against the departure with c for testing model (6.4.1). The first row is for normal error and the second row for chi square error and the third row for uniform error.

6.4.2 Diagnostics with the Wilks Lambda

Consider the linear model as

$$\boldsymbol{Y} = c(\boldsymbol{\beta}_{(1)}\boldsymbol{X}^{(1)}) + (\boldsymbol{\beta}_{(2)}^T\boldsymbol{X}^{(2)}) + \varepsilon \qquad (6.4.2)$$

where $\boldsymbol{Y}$ is q-dimensional, $\boldsymbol{X} = (\boldsymbol{X}^{(1)}, (\boldsymbol{X}^{(2)})^T)^T$ where $\boldsymbol{X}^{(1)}$ is 1-dimensional, $\boldsymbol{X}^{(2)}$ is $(p-1)$-dimensional independently of ε, $\boldsymbol{X}$ is multivariate normal $N(0, \boldsymbol{I}_p)$. Three distributions of the error ε: $N(0, I_q)$, normal; $U_q(-0.5, 0.5)$, uniform on the cube $(-0.5, 0.5)^q$, and $\chi_q^2(1)$, all components following chi-square with degree 1 of freedom respectively. The hypothetical regression function is $\boldsymbol{\beta}_{(2)}^T\boldsymbol{X}^{(2)}$. Therefore the null model corresponds to $c = 0$. In the simulation, we considered $c = 0, 0.1, 0.2, \cdots, 1$ and $p = 3$ and $q = 2$, and

$\boldsymbol{\beta}_{(2)} = (1,1;0,1)$ and $\boldsymbol{\beta}_1 = (1,0)$. Figure 6.3 reports the simulation results. We can find that the Monte Carlo test outperforms the limit one in all of the cases especially when sample size is small.

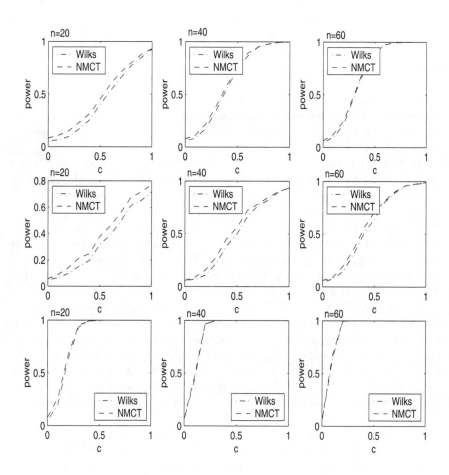

Fig. 6.3. The figure repots the plots of the power against the departure with c for model (6.4.2). The first row is for normal error and the second row for chi square error and the third row for uniform error.

6.4.3 An Application

The 1984 Olympic records data on various track events were collected as reported by Johnson and Wichern (1992). For a relevant dataset of women's track records, Dawkins (1989) used principal component analysis to study the athletic excellence of a given nation and the relative strength of the nation at the various running distances. Zhu (2003) studied the relation between the performance of a nation in long running distance and short running distance. For 55 countries winning times for women's running events at 100, 200, 400, 800, 1,500, 3,000 meters and the Marathon distance are reported in Johnson and Wichern (1992). Now we want to know whether a nation of women whose performance is better in running long distances may also have greater strength at short running distances. To make the analysis more reasonable, the winning time is transformed to speed. Let these speeds be $x_1, \ldots, x_7$. We regard 100, 200 and 400 meters as short running distances, 1,500 meters and longer as long running distances. The hypothetical model is linear by considering the speed of the 100, 200 and 400 meters running events (x_1, x_2, x_3) as the covariates and the speed of the 1,500, 3,000 meters and the Marathon running events (Y_1, Y_2, Y_3) as covariates.

To test the linearity, we used the proposed test $\boldsymbol{TT}_n$ in Section 6.2 and the NMCT in Section 6.3. For NMCT, we assumed two cases: the error follows an elliptically symmetric distribution and a general distribution respectively. Therefore, we used respective algorithms to construct NMCT statistics $\boldsymbol{TT}'_n(U_n)$ and $\boldsymbol{TT}'_n(E_n)$ as reported in Section 6.3. From Figure 6.4, we found that the non-linearity may be mainly from Y_3, the Marathon. There might be quadratic curves in the plots of Y_3 against X_i, $i = 1, 2, 3$. Hence, we chose X_3^2 as a weight function W. With these three tests, the p-values are, respectively, 0.09 with $\boldsymbol{TT}_n$; 0.0001 with $\boldsymbol{TT}'_n(U_n)$ under the elliptical distributional assumption and 0.03 with $\boldsymbol{TT}'_n(U_n)$ under a general distribution assumption. Clearly, Monte Carlo test suggests a rejection for a linear model. Furthermore, since the observation from Figure 6.4 indicates that the non-linearity may be mainly from Y_3, the Marathon and there might be quadratic curves in the plots of Y_3 against X_i, $i = 1, 2, 3$, this implies that the nation with either great strength or weak strength at short running distance may not have good performance in running the Marathon. We then fit a model linearly with X_1, X_2, X_3 and X_3^2. The p-value with the three tests are: 0.97 with $\boldsymbol{TT}_n$; 0.34 with $\boldsymbol{TT}'_n(U_n)$ under the elliptical distribution assumption and 0.99 with $\boldsymbol{TT}'_n(U_n)$ under a general distributional assumption. These tests provide a very strong evidence for the tenability of the model with a two-order polynomial of X_3.

Let us turn to the classical testing problem with likelihood ratio test. First, we note that the speeds of 100 and 200 meters are greatly correlated with correlation coefficient 0.9528. Regressing X_2 on X_1, we obtain $\hat{X}_2 = 1.1492X_1$. Hence, we considered a new model, letting $\tilde{X}_1 = X_2 - 1.1492X_1$, $\tilde{X}_2 = X_2 + 1.1492X_1$,

$$\boldsymbol{Y} = \mathbf{a}\tilde{X}_1 + \mathbf{b}\tilde{X}_2 + \mathbf{d}X_3 + \mathbf{c}X_3^2 + \varepsilon. \tag{6.4.3}$$

The purpose was to test whether $\tilde{X}_1$ has a significant impact for $\boldsymbol{Y}$: that is, the coefficient $a = 0$ or not. The p-values are: 0.08 for the Wilks Lambda; 0.20 for the NMCT with uniformly distributed weights and 0.38 for the NMCT with normally distributed weights. All of the three tests provide evidence that $\tilde{X}_1$ has less impact for $\boldsymbol{Y}$. Hence we can use a model as

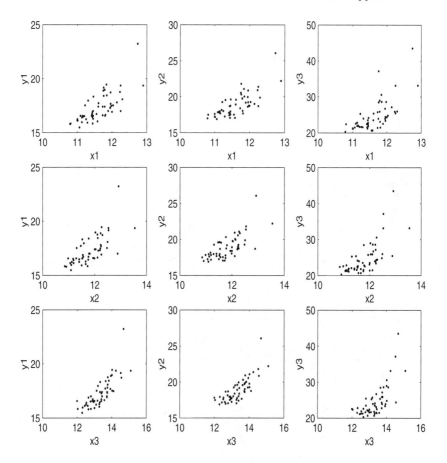

Fig. 6.4. This figure reports the plots of the responses Y_1, Y_2, Y_3 against the covariates X_1, X_2, X_3 for the 1984 Olympic records data.

$$Y = \mathbf{b}\tilde{X}_2 + \mathbf{d}X_3 + \mathbf{c}X_3^2 + \varepsilon$$

to establish the relationship between Y and $\tilde{X}_2, X_3, X_3^2$.

6.5 Appendix

Proof of Lemma 6.2.1. Note that $\boldsymbol{\Sigma}^{-1/2}(\boldsymbol{T} + \boldsymbol{S})$ is normally distributed as $N(\boldsymbol{\Sigma}^{-1/2}\boldsymbol{S}, \boldsymbol{I}_q)$ where $\boldsymbol{I}_q$ is a $q \times q$ identity matrix. Hence, the components, say, u_i $i = 1, \cdots, q$, of $\boldsymbol{\Sigma}^{-1/2}(\boldsymbol{T} + \boldsymbol{S})$ are independent normal with the mean v_i and variance 1 where v_i's are the components of $\boldsymbol{\Sigma}^{-1/2}\boldsymbol{S}$. Therefore, $(\boldsymbol{T} + \boldsymbol{S})^T \boldsymbol{\Sigma}^{-1}(\boldsymbol{T} + \boldsymbol{S})$ can be written as the sum of independent non-central chi-squared variables $\sum_{i=1}^q (u_i + v_i)^2$, each has the non-centrality

v_i^2. From the univariate response case, we know that $P\{(u_i + v_i)^2 \geq c\}$ for any $c > 0$ is smaller when $|v_i|$ gets larger. Consider $q = 2$. Note that the distribution of $(u_1 + v_1)^2 + (u_2 + v_2)^2$ is a convolution of two distributions, each decreasing according to smaller value v_i respectively. First note that $P\{(u_1 + v_1)^2 + (u_2 + v_2)^2 > c\} = \int(1 - F_{1,v_1}(c - x_2))d\,F_{2,v_2}(x_2) = \int(1 - F_{2,v_2}(c - x_1))d\,F_{1,v_1}(x_1)$. Then for any pairs (v_1, v_2) and (v_1', v_2') with $|v_i| \geq |v_i'|$, because of the independence between u_1 and u_2, we derive that

$$P\{(u_1 + v_1)^2 + (u_2 + v_2)^2 > c\} = \int(1 - F_{1,v_1}(c - x_2))d\,F_{2,v_2}(x_2)$$

$$\geq \int(1 - F_{1,v_1'}(c - x_2))d\,F_{2,v_2}(x_2) = \int(1 - F_{2,v_2}(c - x_1))d\,F_{1,v_1'}(x_1)$$

$$\geq \int(1 - F_{2,v_2'}(c - x_2))d\,F_{1,v_1'}(x_2) = P\{(u_1 + v_1')^2 + (u_2 + v_2')^2 > c\}.$$

When we use induction, the same can apply to prove the general case, we omit the details.   □

Proof of Lemma 6.2.2.   Let $\mathbf{V}' = \boldsymbol{\Sigma}^{-1/2}\mathbf{V} =: ((\mathbf{V}')^{(1)}, \cdots, (\mathbf{V}')^{(q)})$. Since $S_i = E(\mathbf{V}^{(i)}\boldsymbol{\eta}^{(i)})$, then $\sum_{i=1}^q v_i^2 = \sum_{i=1}^q E[(\mathbf{V}')^{(i)}\boldsymbol{\eta}^{(i)})]^2$. Invoking the Cauchy-Schwarz inequality and the fact that $E[(\mathbf{V}')^{(i)}]^2 = 1$, we obtain that $\mathbf{S}^T\boldsymbol{\Sigma}^{-1}\mathbf{S} \leq \sum_{i=1}^q(E[\boldsymbol{\eta}^{(i)}])^2$ and the equality holds if and only if $(\mathbf{V}')^{(i)} = \boldsymbol{\eta}^{(i)}/\sqrt{(E(\boldsymbol{\eta}^{(i)})^2)}$.   □

Proof of Theorem 6.3.1.   First note that under the null,

$$\tilde{\mathbf{T}}_n = \frac{1}{\sqrt{n}}\sum_{j=1}^n \mathbf{V}_j \bullet \mathbf{e}_j + o_p(1). \tag{6.5.1}$$

It is easy to see that when the sequence of $\{(\mathbf{x}_i, \|\boldsymbol{\varepsilon}_i\|), i = 1, \ldots, n, \cdots, \}$ is given, $\frac{1}{\sqrt{n}}\sum_{j=1}^n \mathbf{V}_j \bullet \mathbf{e}_j$ has the same distribution as $\frac{1}{\sqrt{n}}\sum_{j=1}^n \mathbf{V}_j \bullet \mathbf{u}_j \bullet \|\mathbf{e}_j\|$ because $\boldsymbol{\varepsilon}_j/\|\boldsymbol{\varepsilon}_j\|$ is independent of $\|\boldsymbol{\varepsilon}_j\|$, and the distribution of $\boldsymbol{\varepsilon}_j/\|\boldsymbol{\varepsilon}_j\|$ is identical to that of $\mathbf{u}_j$ and so do the distributions of the associated unconditional counterparts. This implies that the limit distribution of $\tilde{\mathbf{T}}_n$ is the same as that of $\frac{1}{\sqrt{n}}\sum_{j=1}^n \mathbf{V}_j \bullet \mathbf{u}_j \bullet \|\mathbf{e}_j\|$. Note that this is in turn asymptotically equivalent to $\tilde{\mathbf{T}}_n(\mathbf{U}_n)$. The proof can be done as follows.

Note that in

$$\tilde{\mathbf{T}}_n(\mathbf{U}_n) = \frac{1}{\sqrt{n}}\sum_{j=1}^n \hat{\mathbf{V}}_j \bullet \mathbf{u}_j \bullet \|\hat{\mathbf{e}}_j\|, \tag{6.5.2}$$

$\mathbf{u}_j$ are independent of $\{(\mathbf{x}_j, \mathbf{y}_j), j = 1, \cdots, n\}$, and all estimators involved in $\hat{\mathbf{V}}_j$ and $\hat{\mathbf{e}}_j$ are consistent. Then we can easily derive that, by Taylor expansion,

$$\frac{1}{\sqrt{n}} \sum_{j=1}^{n} \left[\hat{\boldsymbol{V}}_j \bullet \boldsymbol{u}_j \bullet \|\hat{\boldsymbol{e}}_j\| - \boldsymbol{V}_j \bullet \boldsymbol{u}_j \bullet \|\boldsymbol{e}_j\| \right]$$

$$= \frac{1}{\sqrt{n}} \sum_{j=1}^{n} \boldsymbol{u}_j \bullet \left[\hat{\boldsymbol{V}}_j \bullet \|\hat{\boldsymbol{e}}_j\| - \boldsymbol{V}_j \bullet \|\boldsymbol{e}_j\| \right]$$

in probability. The proof is finished. □

Proof of Theorem 6.3.2. To prove the result, we only need to prove two things: asymptotic normality of $\tilde{\mathbf{T}}_n(\boldsymbol{U}_n)$ and the identical of the covariance matrix to the limiting covariance of $\tilde{\boldsymbol{T}}_n$. Note that $\boldsymbol{u}_j$ are independent of $\{(\boldsymbol{x}_j, \boldsymbol{y}_j), j = 1, \cdots, n\}$. Then, when $\{(\boldsymbol{x}_j, \boldsymbol{y}_j)\}$ are given, the covariance matrix of $\tilde{\mathbf{T}}_n(\boldsymbol{U}_n)$ is $(1/n) \sum_{j=1}^{n} (\hat{\boldsymbol{V}}_j \bullet \hat{\boldsymbol{\varepsilon}}_j)(\hat{\boldsymbol{V}}_j \bullet \hat{\boldsymbol{\varepsilon}}_j)^T$. By the consistency of the estimators involved, Taylor expansion and the weak law of large numbers, it is easy to see that this sum converges to $\boldsymbol{E}\left[(\boldsymbol{V} \bullet \boldsymbol{\varepsilon})(\boldsymbol{V} \bullet \boldsymbol{\varepsilon})^T\right]$. This is just the limiting covariance of $\tilde{\boldsymbol{T}}_n$. As for the asymptotic normality, we only need to note that when $\{(\boldsymbol{x}_j, \boldsymbol{y}_j)\}$ are given, $\tilde{\mathbf{T}}_n(\boldsymbol{U}_n)$ is a sum of *i.i.d.* random vectors. combining central limit theorems, we can verify that the limit is distributed as $N(0, \boldsymbol{E}\left[(\boldsymbol{V} \bullet \boldsymbol{\varepsilon})(\boldsymbol{V} \bullet \boldsymbol{\varepsilon})^T\right])$. This is identical to the limit distribution of $\tilde{\boldsymbol{T}}_n$. The proof is finished. □

Proof of Theorem 6.3.3. The proof is almost the same as those for the previous theorems. The details are omitted. □

7

Heteroscedasticity Tests for Regressions

7.1 Introduction

Checking on heteroscedasticity in regression models should be conducted when all of the error terms are assumed to have equal variance. Visual examination of appropriate residual plots is a frequently used technique. The residual analysis has also been applied to constructing formal tests of heteroscedasticity. A rather complete discussion of such tests is given in Carroll and Ruppert (1988, Section 3.4).

When regression and variance functions are parametric, there are some tests in the literature. Examples are score tests (Cook and Weisberg (1983)), quasi-likelihood tests and pseudo-likelihood tests (Davidian and Carroll (1987) and Carroll and Ruppert (1988)). Most of the tests are studied under normality assumption on the distribution of the error. Bickel (1978) extended the classic framework to include "robust" tests in which normality needs not be imposed. Carroll and Ruppert (1981) further studied properties of Bickel's test when the regression function is assumed to be linear.

Recently, some efforts have been devoted to the testing problem with non-parametric regressions. Eubank and Thomas (1993) investigated the case where no parametric form is assumed for the regression function. Their test is a score type test. The input is scalar and the distribution of the error is assumed to be normal. Diblasi and Bowman (1997) recommended a test using non-parametric smoothing techniques in linear model with normal error. The resulting test is actually also a score test with a transformation of the residuals. Again for the case of scalar input, Dette and Munk (1998) proposed a test in the non-parametric regression set-up with fixed design. Their test has a nice property that it does not involve direct estimation of the regression curve and does not depend on the subjective choice of smoothing parameter. In the case of high-dimensional input, Müller and Zhao (1995) proposed a test where, under alternative, the error terms follow a generalized linear model while the regression model is non-parametric. No distributional assumptions are made on the distribution of the error except for some moment properties essentially amounting to existence of the eighth moment. They studied the limit behavior of the test statistic under the null hypothesis, but have not provided theoretical or numerical evidence on the power performance of the test. Eubank and Thomas (1993) investigated the behavior of the test under directional local alternatives.

In this chapter, we shall recommend a unified approach to constructing test statistics which can handle parametric and non-parametric regression models with multi-dimensional input. No assumptions on the distribution of the errors and the form of variance function are made except continuity and fourth moments. Hence the models are more general and the assumptions are weaker than those in Müller and Zhao (1995). Furthermore, regardless of the type of regression functions and the variance functions, the tests can detect the local alternatives distinct $1/\sqrt{n}$ from the null hypothesis. It should be mentioned that although only the results about random designs are presented here, the method can readily handle the cases with fixed designs.

When a non-parametric model is studied, we need to use a non-parametric smoothing method to estimate the regression function, which involves smoothing parameters. It is typically a concern on whether the choice of the smoothing parameters affects seriously estimation efficiency. For our testing problem, the choice will not be crucial. In a wide range of the smoothing parameters, the limit distributions of the test statistics will be identical. We shall show that if the covariable is one-dimensional and the fourth conditional moment of the error given the covariable is a constant function, which is the case when the covariable is independent of the error, the tests will be asymptotically distribution-free. For high-dimensional cases, the tests do not share this property generally. In order to determine critical values, Monte Carlo approximations may be resolutions. We shall investigate the behavior of the classical bootstrap and the NMCT. Related works are Stute, González Manteiga and Presedo Quindimil (1998) and Stute, Thies and Zhu (1998).

7.2 Construction and Properties of Tests

7.2.1 Construction

Consider a regression model:

$$Y = \phi(X) + \varepsilon,$$

where X is a d-dimensional covariate, $E(\varepsilon|X = x) = 0$ and $E(\varepsilon^2|X = x) =: \sigma^2(x)$. Without loss of generality, write them as $E(\varepsilon|x)$ and $E(\varepsilon^2|x)$ respectively. The hypothesis to be tested can be written as, almost surely,

$$H_0 : \quad \sigma^2(\cdot) = \sigma^2, \quad \text{a constant,}$$

versus

$$H_1: \quad \sigma^2(\cdot) \text{ is a non-constant function.}$$

Clearly, if the null hypothesis holds true, σ^2 should just be the variance of ε, $E\varepsilon^2$. Hence H_0 being true is equivalent to, for almost all x,

$$E(\varepsilon^2|x) - E(\varepsilon^2) = 0$$

which is further equivalent to, assuming the distribution function of X, $F(\cdot)$, is continuous,

$$T(x) := \int I(X \leq x)(E(\varepsilon^2|X) - E(\varepsilon^2))dF(X)$$
$$= E(\varepsilon^2(I(X \leq x) - F(x))) = 0, \quad \text{for all } x, \qquad (7.2.1)$$

where "$X \leq x$" means that the components of X are less than or equal to the corresponding components of x. Let $\{(x_1, y_1), \cdots, (x_n, y_n)\}$ be the collected data. The fitted model $\hat{y}_i = \hat{\phi}(x_i)$ based on the data produces the residuals $\hat{\varepsilon}_i = y_i - \hat{y}_i$ where $\hat{\phi}(x)$ is an estimator of $\phi(x)$. The empirical version of $T(x)$ is

$$T_n(x) = \frac{1}{n} \sum_{j=1}^{n} \hat{\varepsilon}_j^2(I(x_j \leq x) - F_n(x)), \qquad (7.2.2)$$

where F_n is the empirical distribution based on x_j's. The test statistic should be a functional of T_n. In this chapter, we define a test of Cramér-von Mises type, the quadratic form of T_n, as

$$W_n = nC_n^{-2} \int [T_n(x)]^2 dF_n(x), \qquad (7.2.3)$$

where C_n^2 is the sample variance of $\hat{\varepsilon}_j^2$, a normalizing constant.

The above procedure can be applied to several types of models. It is possible to apply the idea to some types of model. For different kinds of models, we should estimate regression functions together with their own characters for the consideration of efficiency. In this chapter we only deal with two types of models: that is, parametric and non-parametric regression models.

1. Parametric model: $Y = \phi_\beta(X) + \varepsilon$ where ϕ is a known function. We can estimate β first to get a fitted model $\phi_{\hat{\beta}}(x_j)$ and then the residual $y_j - \phi_{\hat{\beta}}(x_j)$. A test statistic can be defined through (7.2.2) and (7.2.3).

2. Non-parametric model: $Y = \phi(X) + \varepsilon$ where ϕ is an unknown function. Since little is known on the form of ϕ, we have to apply local smoothing technique to estimate it and then to define test.

7.2.2 The Limit Behavior of T_n and W_n

In order to clearly describe the limit behavior of the tests, we give only the details for the linear and pure non-parametric model. Similar results can be derived for parametric models. Suppose that the data $(x_1, y_1), \ldots, (x_n, y_n)$ are *i.i.d.* observations.

Linear Model Case.   The model here is $Y = \alpha + \beta' X + \varepsilon$. The least squares estimator (LSE) of β is defined as, with $\hat{S}$ being the sample covariance matrix of x_j's, $\hat{\beta} = \hat{S}^{-1} \sum_{j=1}^{n} (x_j - \bar{x})(y_j - \bar{y})$. $\hat{\beta}$ is root-n consistent to β, that is $\hat{\beta} - \beta = O_p(1/\sqrt{n})$ provided that the covariance matrix of X, S say, is finite and positive definite. The estimator of α, say $\hat{\alpha}$, can be defined by $\bar{y} - \hat{\beta}'\bar{x}$. Hence $\hat{\varepsilon}_i = y_i - \hat{\alpha} - \hat{\beta}'x_i$. The following theorem states the limit behavior of T_n and then of W_n defined in (7.2.2) and (7.2.3).

THEOREM 7.2.1 *Assume that the distribution, F, of X is continuous and the fourth moments of X and ε are finite and the covariance matrix of X is positive definite. Then under H_0*

$$T_n \Longrightarrow B_1 \qquad (7.2.4)$$

in the Skorohod space $D[-\infty, \infty)^d$ and

$$W_n \Longrightarrow W := C^{-2} \int B_1^2(x)\, d\, F(x),$$

where d is the dimension of X, B_1 is a centered Gaussian process with the covariance function

$$
\begin{aligned}
E(B_1(x)B_1(x_1)) = E\Big[(\varepsilon^2 - \sigma^2)^2 \big(I(X \le x \wedge x_1) - F(x)I(X \le x_1) \\
- F(x_1)I(X \le x) + F(x)F(x_1))\big]\Big] \quad (7.2.5)
\end{aligned}
$$

for all x and x_1 with C^2 being the variance of ε^2 and " $\wedge$ " means taking minimum between two values and if x and x_1 are vectors, the minimum is also a vector whose components are the minimum of corresponding components of the two vectors.

Non-parametric Case.   For non-parametric set-up, the situation becomes more complicated. In the linear model, the estimators of the parameters have the root-n rate of convergence. But this is no longer the case in the non-parametric regression set-up. Typically, the estimator of the regression function may only have $n^{-(m+1)/(2(m+1)+d)}$ rate of convergence to the true regression curve where d is the dimension of the covariate X and m is an indication for the smoothness of the regression function and distribution function of X.(c.f. Subsection 7.5.1 below). We now use the kernel method to define an estimator of the non-parametric function ϕ. Let $g(x) = \phi(x)f(x)$. The kernel estimators of $f(x)$ and $g(x)$ have the following forms:

$$\hat{g}(x) = \frac{1}{n} \sum_{j=1}^{n} y_j K_h(x - x_j),\, \hat{f}(x) = \frac{1}{n} \sum_{j=1}^{n} K_h(x - x_j),$$

$$\hat{\phi}(x) = \hat{g}(x)/\hat{f}(x), \qquad (7.2.6)$$

where h is a bandwidth, $K(\cdot)$ is a kernel function and $\bar{K}_h(\cdot) = \bar{K}(\cdot/h)/h^d$.

THEOREM 7.2.2 *Assume that conditions 1) — 5) listed in Subsection 7.5.1 hold. Then under H_0, the conclusions in Theorem 7.2.1 continue to hold.*

The following consequence is useful in some special cases.

COROLLARY 7.2.1 *In addition to the assumptions in Theorem 7.2.1 or Theorem 7.2.2, assume further under the null hypothesis H_0 that $E((\varepsilon^2 - \sigma^2)^2 | X) = $ a constant. T_n converges weakly to B_1 with the covariance function*

$$E(B_1(x)B_1(x_1)) = C^2 E[(I(X \le x \wedge x_1) - F(x)F(x_1))] \qquad (7.2.7)$$

where C^2 is the variance of ε^2. In the one-dimensional case, W_n converges weakly to $\int_0^1 B^2(x)dx$ with B being the Brownian bridge.

We now investigate how sensitive the test is to alternatives. Consider a sequence of local alternatives indexed by n

$$\sigma^2(x) = \sigma^2 + s(x)/n^a \quad a \ge 0. \qquad (7.2.8)$$

The following conclusion indicates that the test is consistent against all global alternatives (corresponding to $a = 0$) and can detect the local alternatives converging to the null at up to a parametric rate $1/\sqrt{n}$ (corresponding to $0 < a \le 1/2$).

THEOREM 7.2.3 *Assume that the conditions in Theorem 7.2.1 for parametric models or those in Theorem 7.2.2 for non-parametric models hold. Then under the above alternative with $0 \le a < 1/2$, $W_n \to \infty$ as $n \to \infty$ in probability and with $a = 1/2$*

$$T_n \Longrightarrow B_1 + SF,$$

where $SF(x) = E(s(X)(I(X \le x) - F(x))$ is a non-random shift function. Consequently, for a continuous F, $W_n \Longrightarrow C^{-2} \int (B_1 + SF)^2 dF$.

We now discuss the determination of critical values. In the one-dimensional case, if under the null hypothesis the error is independent of X, the critical values can be determined by existing table (e.g. Shorack and Wellner (1986)). But in general, especially in high-dimensional cases, the test statistic W_n may not be distribution-free. Approximations to the null distributions are necessary. In the next section, Monte Carlo approximations are investigated.

7.3 Monte Carlo Approximations

We discuss three Monte Carlo approximations in this section: the Classical bootstrap, the Wild bootstrap and the NMCT.

1. The Classical Bootstrap

a) *Linear Model Case.* Draw b_i's independently from $\hat{\varepsilon}_i$ (with replacement) Let $y_i^* = \hat{\alpha} + \hat{\beta}' x_i + b_i$. The bootstrap LSE of β will be (see Efron (1979))

$$\beta^* = \hat{S}^{-1} \sum_{j=1}^{n} (x_j - \bar{x})(y_j^* - \bar{y}^*), \qquad \alpha^* = \bar{y}^* - (\beta^*)'\bar{x}.$$

The bootstrap residuals are with $b_i^* = b_i - (1/n)\sum_{j=1}^{n} b_j$

$$\hat{\varepsilon}_j^* = y_j^* - \alpha^* - \beta^{*'} x_j = -(\beta^* - \hat{\beta})'(x_j - \bar{x}) + b_j^*$$

which creates a conditional counterpart of T_n

$$T_{n1}^*(x) = \frac{1}{\sqrt{n}} \sum_{j=1}^{n} (\hat{\varepsilon}_j^*)^2 (I(x_j \leq x) - F_n(x))$$

and then

$$W_{n1}^* = C_n^{-2} \int (T_{n1}^*(x))^2 d\, F_n(x). \qquad (7.3.1)$$

b) *Non-parametric Model Case.* Let e_i^* be independently drawn from $\hat{\varepsilon}_i = y_i - \hat{\phi}(x_i)$ (with replacement). Define a bootstrap process by

$$T_{n1}^*(x) = \frac{1}{\sqrt{n}} \sum_{j=1}^{n} (e_j^*)^2 (I(x_j \leq x) - F_n(x))$$

where $g^*(x) = \frac{1}{n}\sum_{j=1}^{n} e_j^* K_h(x - x_j)/\hat{f}(x)$. Define a bootstrap statistic as

$$W_{n1}^* = C_n^{-2} \int (T_{n1}^*(x))^2 d\, F_n(x). \qquad (7.3.2)$$

THEOREM 7.3.1 *Under the conditions in Theorem 7.2.1 or Theorem 7.2.2, for almost all sequences $\{(x_1, y_1), \cdots, (x_n, y_n), \cdots\}$, the limit conditional distribution of W_{n1}^* in (7.3.1) or (7.3.2) is identical to the limit null distribution of W_n with the covariance function in Corollary 7.2.1.*

REMARK 7.3.1 *From this theorem, we learn that, no matter whether the null hypothesis holds true, the limit of the bootstrap test statistic W_{n1}^* in (7.3.1) or (7.3.2) is only the same as that of W_n when $E(\varepsilon^4|X)$ is a constant under H_0. Although this is the case when, under H_0, the covariate and the error are independent, the bootstrap test statistic is inconsistent in general. Consequently, the classical bootstrap is only applicable in some special cases.*

2. The Wild Bootstrap

In our case, the wild bootstrap, as in Wu (1986) and Härdle and Mammen (1993), does not work. To see this, consider the linear model and define $x_i^* = x_i$, $y_i^* = \hat{\beta}'x_i + \varepsilon_i^*$ with $\varepsilon_i^* = w_i\hat{\varepsilon}_i$ where w_i are $i.i.d.$ artificial bounded variables generated by computer such that

$$Ew_i = 0, \qquad \text{Var}(w_i) = 1.$$

By the least squares method, the bootstrap LSE of β based on the data $\{(x_1^*, y_1^*), \cdots, (x_n^*, y_n^*)\}$, say $\hat{\beta}^*$, satisfies $\hat{\beta}^* - \hat{\beta} = \hat{\beta} - \beta + O_p(1/\sqrt{n})$ almost surely, see Liu (1988) or Härdle and Mammen (1993). The bootstrap residuals will be $y_i^* - (\hat{\beta}^*)'x_i = \varepsilon_i^* - (\hat{\beta}^* - \hat{\beta})'x_i$. The bootstrap version of T_n, say T_n^*, is then

$$T_n^*(x) = \frac{1}{\sqrt{n}} \sum_{j=1}^{n} (y_i^* - (\hat{\beta}^*)'x_i)^2 (I(x_j \leq x) - F_n(x))$$

$$= \frac{1}{\sqrt{n}} \sum_{j=1}^{n} ((y_i^* - (\hat{\beta}^*)'x_i)^2 - \hat{\sigma}^2)(I(x_j \leq x) - F_n(x))$$

which has a similar decomposition to T_n:

$$\frac{1}{\sqrt{n}} \sum_{j=1}^{n} (w_i^2 \hat{\varepsilon}_j^2 - \hat{\sigma}^2)(I(x_j \leq x) - F_n(x))$$

$$-(\hat{\beta}^* - \hat{\beta})' \frac{2}{\sqrt{n}} \sum_{j=1}^{n} w_i \hat{\varepsilon}_j x_j (I(x_j \leq x) - F_n(x))$$

$$+\sqrt{n}(\hat{\beta}^* - \hat{\beta})' \frac{1}{n} \sum_{j=1}^{n} x_j x_j' (I(x_j \leq x) - F_n(x))(\hat{\beta}^* - \hat{\beta})$$

$$=: I_1^*(x) - I_2^*(x) + I_3^*(x). \tag{7.3.3}$$

The following is an outline of the proof. Intuitively, note that due to the root-n consistency of $\hat{\beta}^*$ the process $I_3^* = \{I_3^*(x), x \in R^d\}$ is clearly converging to zero in distribution. For the process $I_2^* = \{I_2^*(x), x \in R^d\}$, the sum term is a centered process which has a finite limit, and, together with the root-n consistency of $\hat{\beta}^*$, I_2^* goes to zero. When $w_i^2 = 1$, the process $I_1^* = \{I_1^*(x), x \in R^d\}$ is simply equal to T_n which is a non-random function for a given data and when w_i^2 is not a constant, I_1^* can be verified to converge weakly to a different limit from the limit of T_n. Hence asymptotically, the conditional counterpart of W_n, say W_n^*, is inconsistent.

3. The NMCT

We consider a simple case. Let $e_1, \ldots, e_n$ be $i.i.d.$ variables equally likely taking values ± 1. The NMCT counterpart of T_n is defined by

$$T_n(E_n, x) = \frac{1}{\sqrt{n}} \sum_{j=1}^{n} e_i(\hat{\varepsilon}_i^2 - \hat{\sigma}^2)(I(x_j \leq x) - F_n(x)) \qquad (7.3.4)$$

where $E_n := (e_1, \cdots, e_n)$. The resulting test statistic is

$$W_n(E_n) = C_n^{-2} \int (T_n(E_n, x))^2 d\, F_n(x)). \qquad (7.3.5)$$

Generate m sets of E_n, say $E_n^{(i)}, i = 1, ..., m$ and then get m values of $W_n(E_n)$, say $W_n(E_n^{(i)}), i = 1, ..., m$. The $1 - \alpha$ quantile of $W_n(E_n^{(i)})$'s will be as the α-level critical value for W_n.

THEOREM 7.3.2 *Assume that the conditions in Theorem 7.2.1 for parametric models or those of Theorem 7.2.2 for non-parametric models hold. When $\sigma^2(x) = \sigma^2 + s(x)/n^a$ for some $a > 0$ for almost all sequences $\{(x_1, y_1), \cdots, (x_n, y_n), \cdots\}$ the conditional distribution of $W_n(E_n)$ converges to the limiting null distribution of W_n in Theorem 7.2.1.*

As $s(\cdot) = 0$ corresponds to the null hypothesis and $s(\cdot) \neq 0$ to the local alternative, this conclusion indicates that the critical values determined by the NMCT, under local alternatives, equal approximately the ones under the null hypothesis. Hence the critical values remain unaffected in the large sample sense by the underlying model with small perturbations from constant conditional variance.

For global alternatives, that is, $a = 0$, $W_n(E_n)$ has a finite limit while W_n goes to infinity. Therefore the test is also consistent.

7.4 A Simulation Study

To give evidence of how the proposed tests work, a small sample simulation experiment was conducted. Comparisons were made with some existing tests in the literature.

Following Diblasi and Bowman (1997), the variance functions were $\sigma_1(x) = 1$, $\sigma_2(x) = 0.25 + x$, $\sigma_3(x) = 0.25 + 4(x - 0.25)^2$ and $\sigma_4(x) = 0.25 \exp(x \log(5))$.

Throughout, the basic experiment was performed 500 times and the nominal level was 0.05. The tables below show the percents of times out of 500 that each procedure rejected the null hypothesis. Bandwidth selection is a concern. But similar to Eubank and Thomas (1993) who used a spline smoother, the difficulty with the theorems' treatment of the bandwidth is that it does not allow the data-driven choices for h. In homoscedastic cases, the generalized cross validation (GCV) method works well for estimating the bandwidth while it may not be useful with the heteroscedastic errors. As Diblasi and Bowman (1997) did, we considered several values of h in a fairly wide range and empirically chose it. See Tables 6.1 and 6.3 and the discussions below. We also calculated the average value of 500 data-driven bandwidths obtained by GCV under the null hypothesis to help our choice. In the simulation with model (7.4.1) below, as Diblasi and Bowman (1997) used, we reported the empirical powers of the tests with $h = 0.08, 0.16$ and 0.32.

In the tables, the row "Linear" means that we used the least square method to estimate the parameters in the model. "Nonpara" means that we regarded the underlying model as a non-parametric one and estimate the regression function by the kernel method as Diblasi and Bowman (1997). The kernel function was $(15/16)(1-t^2)^2 I(t^2 \leq 1)$ (see, e.g., Härdle and Mammen (1993)). We considered an one-dimensional model as

$$y_i = \beta_0 + \beta_1 x_i + \varepsilon_i, \quad i = 1, \cdots, n, \tag{7.4.1}$$

where the ε_i were the independent normal random variables with zero means and x_i has a uniform distribution on [0,1]. In the simulation, the values for the parameters were $\beta_0 = 1$ and $\beta_1 = 2$.

The critical value of our asymptotic test is given e.g. on page 748 of Shorack and Wellner (1986). In Tables 6.1 and 6.2, we denote that C&W : Cook & Weisberg's test in Cook and Weisberg (1983), D&B : the bootstrap test in Diblasi and Bowman (1997), NEW: the asymptotic test in the paper, NMCT: the NMCT, CBT : the classical Bootstrap test.

Table 6.1. Empirical sizes of the tests ($\sigma_1(x) = 1$)

$n = 50$	C&W	D&B	New	NMCT	CBT
Linear	4.4		5.0	6.4	5.6
Nonpara. $h = 0.08$		4.6	6.0	7.6	9.2
Nonpara. $h = 0.16$		4.4	5.0	5.8	6.4
Nonpara. $h = 0.32$		4.8	5.0	4.4	4.8
$n = 70$	C&W	D&B	New	NMCT	CBT
Linear	5.0		3.8	4.6	5.0
Nonpara. $h = 0.08$		4.6	5.4	6.6	8.2
Nonpara. $h = 0.16$		4.2	5.6	5.8	6.8
Nonpara. $h = 0.32$		5.2	4.4	4.6	5.2

In view of Table 6.1, one may see that the asymptotic test has the best performance among the tests on holding the level of significance. In a fairly wide range of bandwidth values, the size of the asymptotic test remains quite close to the nominal level. This fits well with the theoretical assertions in the theorems. Not surprisingly, the C&W test also has good performance in a normal linear model. For the D&B test and the other two conditional tests, properly choosing the bandwidth is needed, but one may see that the performance of the conditional tests is good at $h = 0.32$ and is not bad at

$h = 0.16$ either. The average values of GCV bandwidths were $\bar{h}_{GCV} = 0.30$ and $\bar{h}_{GCV} = 0.28$ according to $n = 50$ and $n = 70$ respectively.

For the powers presented in Table 6.2, we chose $h = 0.32$ in the non-parametric cases. The comparison with the parametric C&W test whose assumption for the variance of the errors is $\sigma_0^2 \exp(\lambda x_i), i = 1, \ldots, n$, where λ is an unknown parameter. The scores were selected as the centered derivative of the variance function as Cook and Weisberg (1983) did. The power performances of tests rested on the variance functions. When the parametric assumption is appropriate, the C&W test has the best performance. But there is no a single test which outperforms others in all cases. Consequently, the asymptotic test can be recommended because it needs less computational work and can be applied to various models.

Table 6.2. Empirical powers of the tests

$n = 50$	C&W	D&B	New	NMCT	CBT
$\sigma_2(x) = 0.25 + x$					
Linear	92.2		85.4	91.4	87.0
Nonpara.		84.8	85.4	87.0	85.4
$\sigma_3(x) = 0.25 + 4(x - 0.25)^2$					
Linear	37.6		41.4	19.2	44.8
Nonpara.		83.2	39.4	17.2	45.8
$\sigma_4(x) = 0.25 \exp(x \log 5)$					
Linear	97.0		93.6	96.4	94.4
Nonpara.		90.6	92.4	91.6	93.0
$n = 70$	C&W	D&B	New	NMCT	CBT
$\sigma_2(x) = 0.25 + x$					
Linear	97.0		95.4	96.4	96.0
Nonpara.		94.4	94.2	94.8	94.6
$\sigma_3(x) = 0.25 + 4(x - 0.25)^2$					
Linear	40.8		59.8	23.0	60.8
Nonpara.		96.4	59.8	21.6	59.6
$\sigma_4(x) = 0.25 \exp(x \log 5)$					
Linear	98.8		99.2	99.4	99.0
Nonpara.		98.4	98.6	98.2	98.2

We also performed another set of simulations. The underlying model was

$$y_i = \beta_0 + \beta_1 x_{i1} + \beta_2 x_{i2} + \varepsilon_i, \quad i = 1, \cdots, n. \tag{7.4.2}$$

In the simulation, $\beta_0 = 1$, $\beta_1 = 2$ and $\beta_2 = 3$ were used and ε_i were normal variables with zero mean. We generated x_{i1} and x_{i2} in the following way: In the interval $[0,1]$, let $\tilde{x}_{i1}$ be from the uniform distribution and $\tilde{x}_{i2} = (2\tilde{x}_{i1})^2$. Define $x_{i1} = \tilde{x}_{i1} + 0.5\tilde{x}_{i2}$ and $x_{i2} = 0.5\tilde{x}_{i1} + \tilde{x}_{i2}$. Hence (7.4.2) is a model associated with one-dimensional variable $\tilde{x}_{i1}$. The asymptotic test is still available. The performance of the re-sampling tests were also investigated. Similar to Diblasi and Bowman (1997), $h = 0.16, 0.32$ and $= 0.64$ were used. The average values of 500 bandwidths obtained by GCV were $\bar{h}_{GCV} = 0.38$ and $\bar{h}_{GCV} = 0.34$ according to $n = 50$ and $n = 70$ respectively.

Based on Table 6.3, the performance of the asymptotic test is still encouraging.

Table 6.3. Empirical sizes of the tests $(\sigma_1(x) = 1)$

$n = 50$	C&W	NEW	NMCT	CBT
Linear	4.6	5.0	6.4	5.8
Nonpara. $h = 0.16$		6.4	9.8	16.2
Nonpara. $h = 0.32$		5.2	4.4	6.2
Nonpara. $h = 0.64$		4.8	1.8	4.0
$n = 70$	C&W	NEW	NMCT	CBT
Linear	5.8	5.2	6.8	5.8
Nonpara. $h = 0.16$		6.6	10.2	14.6
Nonpara. $h = 0.32$		5.4	4.2	6.4
Nonpara. $h = 0.64$		5.0	3.6	6.0

Table 6.4. Empirical powers of the tests

$n = 50$	C&W	NEW	NMCT	CBT
$\sigma_2(x) = 0.25 + x_1$	$n = 50$			
Linear	78.4	82.0	82.4	73.8
Nonpara. $h = 0.32$		81.4	80.8	77.2
$\sigma_3(x) = 0.25 + 4(x_1 - 0.25)^2$				
Linear	92.4	99.8	99.6	97.8
Nonpara. $h = 0.32$		99.8	99.8	99.2
$\sigma_4(x) = 0.25 \exp(x_1 \log 5)$				
Linear	95.2	98.4	99.8	94.4
Nonpara. $h = 0.32$		98.6	99.8	98.2
$n = 70$	C&W	NEW	WBT	CBT
$\sigma_2(x) = 0.25 + x_1$				
Linear	90.6	93.2	93.2	89.2
Nonpara. $h = 0.32$		92.0	92.4	91.8
$\sigma_3(x) = 0.25 + 4(x_1 - 0.25)^2$				
Linear	94.4	100.0	99.8	99.8
Nonpara. $h = 0.32$		100.0	100.0	100.0
$\sigma_4(x) = 0.25 \exp(x_1 \log 5)$				
Linear	97.0	99.4	99.6	96.4
Nonpara. $h = 0.32$		100.0	99.4	98.6

For the powers in Table 6.4, $h = 0.32$ was considered. The asymptotic test and the wild bootstrap test are superior over the others. The scores of C&W test were selected as the centered derivative of the variance function

$\sigma_0^2 \exp(\lambda x_1)$. It is a bit surprising that the score test cannot outperform other tests, even if the parametric assumption is appropriate. But looking at the results reported in Table 1 of Eubank and Thomas (1993), we learnt that the derivative of the variance function might not always be the optimal choice. Note that the test in Eubank and Thomas (1993) is, in univariate cases, a generalization of C & W test. In their simulation, the test T_1 cannot do better than the test T_2 even in the case in which the parametric assumption for the variance of the errors is appropriate for T_1.

Furthermore, it is noted that the test of Eubank and Thomas (1993) is also asymptotically distribution-free in the case where the covariate is one-dimensional and the error is normally distributed. The limiting null distribution is χ^2- distribution with 1 degree of freedom. Their test is defined by a quadratic form of the weighted sum of the squares of the residuals. It is also a Cramér-von Mises type. The choice of weights is involved. For obtaining good power performance, the weights are related to alternatives. In their simulation, three choices of weights were used. The corresponding tests were written as T_1, T_2 and T_3. Following them, we reported the percents of times out of 1000 that each procedure rejected the null hypothesis for the four variance functions: $v_0(x) = 1, v_1(x) = \exp(x), v_2(x) = \exp(x^2)$ and $v_3(x) = 2$ for $x > 0.5$ and $v_3(x) = 1$ for $x < 0.5$. Sample sizes were $n = 100$ and $n = 200$. In the simulation, two regression functions were considered: $\phi_1(x) = 3 + 2.7x + 3x^2$ and $\phi_2(x) = 3exp(-2x)$. Similar to the tests T_1 and T_2 on page 149 in Eubank and Thomas (1993), our asymptotic test is also reasonably insensitive to the choice of regression function ϕ, we in Table 6.5 only report the results associated with ϕ_1. The bandwidth was $h = 0.32$.

Table 6.5. Empirical powers of the asymptotic test

variance	v_0	v_1	v_2	v_3
$n = 100$	5.4	46.0	46.3	51.0
$n = 200$	5.0	76.5	78.0	82.0

Our asymptotic test can hold the level well, while as noted in Eubank and Thomas (1993), their tests have some difficulty on holding the significance level especially for $n = 100$. Their T_2 is more sensitive to alternative than ours in the cases conducted.

To summarize, the asymptotic test may be a good choice in the case that critical values can be determined, otherwise, the NMCT is worthwhile to recommend.

7.5 Proofs of the theorems

7.5.1 A Set of Conditions

The following conditions are required for the theorems in Sections 7.2 and 7.3

1). For non-negative integers $m_1, \ldots, m_d$, $m - m_i + \cdots + m_d$, $g(x) = \phi(x)f(x)$ and $f(x)$ are m_i-times differentiable about the i-th component of x, say x_i, and their m-th derivatives of g and f, say $g^{(m)}(x)$, and $f^{(m)}(x)$, satisfy the following condition, that is, there exists a neighborhood of the origin, say U, and a constant $c > 0$ such that for any $u \in U$ and some $0 < r < 1$

$$|f^{(m)}(x+u) - f^{(m)}(x)| \le c|u|,$$
$$|g^{(m)}(x+u) - g^{(m)}(x)| \le c|u|.$$

2). $E|Y|^4 < \infty$ and $E|X|^4 < \infty$.

3). The continuous kernel function $K(\cdot) = \prod_{i=1}^{d} K^{(i)}(\cdot)$ satisfies the following properties:

   a) the support of $K^{(i)}(\cdot)$ is the interval $[-1, 1]$ for $i = 1, \ldots, d$;

   b) $K^{(i)}(\cdot)$ is symmetric about 0;

   c) $\int_{-1}^{1} K^{(i)}(u)du = 1$, and $\int_{-1}^{1} u^l K^{(i)}(u)du = 0, l = 0, \ldots m_i, i = 1, \ldots d.$

4). As $n \to \infty$ $h \sim n^{-c_1}$ where the positive numbers c_1 satisfies that $\frac{1}{4(m+1)} < c_1 < \frac{1}{2d}$ for $d < 2(m+1)$, where the notation "$\sim$" means that two quantities have the same convergence order.

5). $0 < c_1 \le \inf f(x) \le \sup f(x) \le c_2 < \infty.$

REMARK 7.5.1 *Conditions 1) is concerned with the smoothness of the density function of x and the regression curve $\phi(x)$. Without further restriction on the regression curve, Condition 2) is necessary for the asymptotic convergence of T_n and then of W_n. We note that the asymptotic behavior of T_n in Theorem 7.2.2 does not depend on the choice of the bandwidth h when Conditions 4) is fulfilled. The choice of h is relatively flexible. For example in the one-dimensional cases, $d = 1$ and $m = 0$, the range of h is near from $n^{-\frac{1}{2}}$ to $n^{-\frac{1}{r}}$ which contains the optimal convergence rate of $h = O_p(n^{-\frac{1}{3}})$. Hence it may be said that the test W_n is not sensitive for the choice of the smoothing parameter h. Condition 5) is a typical restriction that avoids boundary effect problem.*

Since similar arguments can be applied to develop proofs for multivariate cases, we only investigate the univariate situation here.

7.5.2 Proofs of the Theorems in Section 2

Proof of Theorem 7.2.1.   Since the x_j are assumed to be scalar as mentioned above, the notations will be simpler. First it is known that $\hat{S} \to S$ in probability and $\sup_x |F_n(x) - F(x)| = O_p(1/\sqrt{n})$. Noticing that $\hat{\varepsilon}_j = (\epsilon_j - \bar{\varepsilon}) - (\hat{\beta} - \beta)(x_j - \bar{x})$, $\hat{\varepsilon}_j^2 = (\epsilon_j - \bar{\varepsilon})^2 - 2(\hat{\beta} - \beta)(x_j - \bar{x})(\varepsilon_j - \bar{\varepsilon}) + (\hat{\beta} - \beta)^2(x_j - \bar{x})^2$, and $(\bar{x} - EX) = O_p(1/\sqrt{n})$, $\bar{\varepsilon} = O_p(1/\sqrt{n})$, we then have

$$T_n(x) = \frac{1}{\sqrt{n}} \sum_{j=1}^{n} (\varepsilon_j^2 - \sigma^2)(I(x_j \le x) - F(x))$$

$$-\frac{2}{\sqrt{n}} (\hat{\beta} - \beta) \sum_{j=1}^{n} \varepsilon_j (x_j - EX)(I(x_j \le x) - F(x))$$

$$+\frac{1}{\sqrt{n}} (\hat{\beta} - \beta)^2 \sum_{j=1}^{n} (x_j - EX)^2 (I(x_j \le x) - F(x)) + O_p(1/\sqrt{n})$$

$$=: I_1(x) - I_2(x) + I_3(x) + O_p(1/\sqrt{n}). \tag{7.5.1}$$

Since the class of functions $f_x(X) = X^2(I(X \le x) - F(x))$ over all indices x is a VC class (see, e.g. Pollard (1984), Giné and Zinn (1984)), $(1/n)\sum_{j=1}^{n} x_j^2(I(x_j \le x) - F(x)) \to E(X^2(I(X \le x) - F(x)))$ a.s. uniformly on x (see Pollard (1984), p.25). Then $I_3(x) = O_p(1/\sqrt{n})$ a.s. uniformly on x. Again the class of functions $f_{1x}(X) = X\varepsilon(I(X \le x) - F(x))$ for all indices x is also a VC class, hence the Equicontinuity lemma (Pollard (1984), p. 150) holds true. By Theorem VII 21 (Pollard (1984), p. 157) $1/\sqrt{n}\sum_{j=1}^{n} \varepsilon_j x_j I(x_j \le x)$ converges weakly to a centered Gaussian process. This implies that, combining with (7.5.1), $I_2(x) = O_p(1/\sqrt{n})$ a.s. uniformly on x. Applying Theorem VII 21 of Pollard (1984) again, I_1 converges weakly to the process B_1 defined in (7.2.5) upon noticing that $C_n \to C$ in probability. From this conclusion, we immediately get that under the conditions in Theorem 7.2.1, applying the continuous mapping theorem, W_n converges weakly to W. This completes the proof. $\square$

Proof of Theorem 7.2.2. Similar to the decomposition of T_n in (7.5.1), we have

$$T_n(x) = \frac{1}{\sqrt{n}} \sum_{j=1}^{n} (\varepsilon_j^2 - \sigma^2)(I(x_j \le x) - F(x))$$

$$-\frac{2}{\sqrt{n}} \sum_{j=1}^{n} \varepsilon_j (\hat{\phi}(x_j) - \phi(x_j))(I(x_j \le x) - F(x))$$

$$+\frac{1}{\sqrt{n}} \sum_{j=1}^{n} (\hat{\phi}(x_j) - \phi(x_j))^2 (I(x_j \le x) - F(x)) + o_p(1)$$

$$=: I_4(x) - I_5(x) + I_6(x) + o_p(1). \tag{7.5.2}$$

What we now need to do is to show that I_5 and I_6 tend to zero in probability. For I_6, we have that

$$\sup_x |I_6(x)| \le \frac{1}{\sqrt{n}} \sum_{j=1}^{n} (\hat{\phi}(x_j) - \phi(x_j))^2$$

which is $\sqrt{n}$ multiplying the mean square error of the estimator of ϕ. Under conditions assumed in Subsection 7.5.1, it is easy to see that the mean square error of the estimator of ϕ is $O_p((1/\sqrt{nh}+h^{m+1})^2)$. We are then able to show that $I_6(x) = O_p(1/\sqrt{nh} + \sqrt{n}h^{2(m+1)})$ uniformly on x.

We now prove that I_5 converges to zero in probability uniformly on x. It can be derived that

$$
\begin{aligned}
\hat{\phi}(x) - \phi(x) &= \frac{\hat{g}(x)}{\hat{f}(x)} - \frac{g(x)}{f(x)} \\
&= \frac{\hat{g}(x) - g(x)}{f(x)} - \phi(x)\frac{\hat{f}(x) - f(x)}{f(x)} \\
&\quad - \frac{(\hat{g}(x) - g(x))(\hat{f}(x) - f(x))}{\hat{f}(x)f(x)} \\
&\quad + \frac{\phi(x)(\hat{f}(x) - f(x))^2}{\hat{f}(x)f(x)}.
\end{aligned}
\tag{7.5.3}
$$

By the similar arguments for computing mean square error of estimator, we have that, along with conditions 1) through 3),

$$
\sum_{j=1}^{n} \left(\hat{g}(x) - g(x)\right)^2 = O_p((\log n)^4/(nh) + h^{2(m+1)}),
$$

$$
\sum_{j=1}^{n} \left(\hat{f}(x) - f(x)\right)^2 = O_p((\log n)^4/(nh) + h^{2(m+1)}).
$$

Combining (7.5.3) and condition 5)

$$
\begin{aligned}
I_5(x) &= \frac{2}{\sqrt{n}} \sum_{j=1}^{n} \varepsilon_j \left(\frac{\hat{g}(x_j) - g(x_j)}{f(x_j)}\right)(I(x_j \le x) - F(x)) \\
&\quad - \frac{1}{\sqrt{n}} \sum_{j=1}^{n} \varepsilon_j \left(\phi(x_j)\frac{\hat{f}(x_j) - f(x_j)}{f(x)}\right)(I(x_j \le x) - F(x)) \\
&\quad + O_p((\log n)^4/\sqrt{n}h + \sqrt{n}h^{2(m+1)}) \\
&=: J_1(x) - J_2(x) + O_p((\log n)^4/\sqrt{n}h + \sqrt{n}h^{2(m+1)}).
\end{aligned}
\tag{7.5.4}
$$

We shall now prove that J_1 and J_2 converge to zero in probability uniformly on x. Since the arguments are similar for both cases, we give details only for J_1.

Let

$$
W_h^1(x_i, x_j, y_i, \varepsilon_j, x) = \frac{(y_i K((x_i - x_j)/h) - g(x_j))\phi(x_j)\varepsilon_j}{hf(x_j)}(I(x_j \le x) - F(x))
$$

and observe that

$$J_1(x) = \frac{1}{2n^{3/2}} \sum_{i \neq j}^{n} W_h^1(x_i, x_j, y_i, \varepsilon_j, x) + W_h^1(x_j, x_i, y_j, \varepsilon_i, x) + O_p(\frac{1}{\sqrt{nh}})$$

$$=: J_1'(x) + O_p(\frac{1}{\sqrt{nh}}) \tag{7.5.5}$$

Let $\eta = (X, Y, \varepsilon)$. Define

$$W_h(\eta_i, \eta_j, x) = h\Big(W_h^1(\eta_i, \eta_j, x) + W_h^1(\eta_j, \eta_i, x)\Big)$$
$$- h\Big(E(W_h^1(\eta, \eta_j, x)|\eta_j)) + E(W_h^1(\eta_i, \eta, x)|\eta_i))\Big)$$

where $E(W_h^1(\eta, \eta_j, x)|\eta_j))$ is the conditional expectation of W_h^1 given η_j and

$$J_1'' = \frac{1}{2n^{3/2}} \sum_{i \neq j}^{n} W_h(\eta_i, \eta_j, x).$$

Hence J_1'' is a U-process (see, e.g. Nolan and Pollard (1987)). As is well known, the class of indicator functions is a VC class. Note that for any x the function $W_h^1(\cdot, x)$ is the product of the centered indicator function $(I(\cdot \leq x) - F(x))$ and a given function which is independent of x. We then have that for any fixed n the class of functions $\mathcal{G}_n = \{W_h(\cdot, x) : x \in R^1\}$ is a vector space of real functions having the same dimension as that consisting of indicator functions. Therefore, by Lemma II.18 (Pollard (1984), p. 20), for any fixed n, $\mathcal{G}_n$ is the VC class whose degree is not greater than that of the class of indicator functions. Note that $E(W_h(\eta_1, \eta_2, x) = 0$. Therefore $\mathcal{G}_n$ is P-degenerate with envelope

$$G_n(\eta_1, \eta_2)$$
$$= \left| \frac{(y_1 K(\frac{(x_1-x_2)}{h})-g(x_2))\phi(x_2)\varepsilon_2}{f(x_2)} \right| + \left| \frac{(y_2 K(\frac{(x_2-x_1)}{h})-g(x_1))\phi(x_1)\varepsilon_1}{f(x_1)} \right|.$$

By Theorem 6 of Nolan and Pollard (1987) on p. 786, we have

$$E \sup_x |\sum_{i, j} W_h(\eta_i, \eta_j, x)| \leq cE(\alpha_n + \gamma_n J_n(\theta_n/\gamma_n))$$

$$J_n(s) = \int_0^s \log N_2(u, T_n, \mathcal{G}_n, G_n) du,$$

$$\gamma_n = (T_n G_n^2)^{1/2}, \qquad \alpha_n = \frac{1}{4} \sup_{g \in \mathcal{G}_n} (T_n g^2)^{1/2},$$

$$T_n g^2 := \sum_{i \neq j} g^2(\eta_{2i}, \eta_{2j}) + g^2(\eta_{2i}, \eta_{2j-1}) + g^2(\eta_{2i-1}, \eta_{2j}) + g^2(\eta_{2i-1}, \eta_{2j-1})$$

and $N_2(\cdot, T_n, \mathcal{G}_n, G_n)$ is the covering number of $\mathcal{G}_n$ under L_2 metric with the measure T_n and the envelope G_n. As $\mathcal{G}_n$ is the VC class, following the argument of Approximation lemma II 2.25 (Pollard (1984), p. 27) the covering number $N_2(uT_n/n^2G_n^2, T_n/n^2, \mathcal{G}_n, G_n)$ can be bounded by cu^{-w_1} for some positive c and w_1, both being independent of n and T_n. Further in probability for large n

$$T_n G_n^2 \leq 2 \sum_{j=1}^{n} \sum_{i=1}^{n} \left(\frac{(y_i K((x_i - x_j)/h) - g(x_j))\phi(x_j)\varepsilon_j}{f(x_j)} \right)^2$$

$$= O(hn^2 \log^2 n) \quad a.s.$$

Hence for large n, $T_n/n^2 G_n^2$ is smaller than 1 and $N_2(u, T_n/n^2, \mathcal{G}_n, G_n) \leq cu^{-w_1}$. Note that $N_2(u, T_n, \mathcal{G}_n, G_n) = N_2(u/n^2, T_n/n^2, \mathcal{G}_n, G_n)$. We can then derive that

$$J_n(\theta_n/\gamma_n) \leq J_n(1/4)$$

$$= n^2 \int_0^{1/(4n^2)} \log N_2(u, T_n/n^2, \mathcal{G}_1, G) d\,u$$

$$= -cn^2 \int_0^{1/(4n^2)} \log u\, d\,u = c \log n$$

and

$$\gamma_n^2 = T_n G_n^2 = O(hn^2 \log^2 n) \quad a.s.$$

Therefore for large n, $E \sup_x |\sum_{i,\,j} W_h(\eta_i, \eta_j, x)| \leq c\sqrt{hn} \log n$. This yields that $E \sup_x |J_1''(x)| \leq c\sqrt{h} \log n/\sqrt{n}$, and then

$$J_1'' = h\frac{1}{\sqrt{n}} \sum_{j=1}^{n} E(W_h^1(\eta, \eta_j, x)|\eta_j) + O_p(\sqrt{h} \log n/\sqrt{n}), \qquad (7.5.6)$$

equivalently

$$J_1' = \frac{1}{\sqrt{n}} \sum_{j=1}^{n} E(W_h^1(\eta, \eta_j, x)|\eta_j) + O_p(\log n/\sqrt{nh})$$

$$= \frac{1}{\sqrt{n}} \sum_{j=1}^{n} E\left((Y - g(x_j))K_h(X - x_j)\phi(x_j)(I(x_j \leq x) - F(x))\frac{\varepsilon_j}{f(x_j)} \right)$$

$$+ O_p(\log n/\sqrt{nh})$$

$$=: J_3(x) + O_p(\log n/\sqrt{nh}). \qquad (7.5.7)$$

By conditions 1) and 3) in Subsection 7.5.1, for each x_j

$$E(Y - g(x_j))K_h(X - x_j) = E(g(X) - g(x_j))K_h(X - x_j)$$

$$= E(g(x_j + hu) - g(x_j))K(u) = O(h^m)$$

and

$$\mathrm{Var}(E((Y - g(x_j))K_h(X - x_j)\phi(x_j)(I(x_j \le x) - F(x))\frac{\varepsilon_j}{f(x_j)})))$$
$$= O(h^{2(m+1)}).$$

Applying Theorem 3.1 of Zhu (1993) or mimicking the proof of Theorem II.37 (Pollard (1984), p. 34) gives

$$\sup_x |J_3(x)| = o(h^{2(m+1)}(\log n)^2) \qquad a.s.$$

The proof is concluded from combining (7.5.5), (7.5.7) with condition 4.   □

Proof of Corollary 7.2.1. When $E((\varepsilon^2 - \sigma^2)^2|X)$ equals a constant, it should be C^2. Making a time transformation, we can derive that $W = \int_0^1 B(x)^2 dx$ with B being a Brownian bridge on [0,1].   □

Proof of Theorem 7.2.3. It is easy to see that either (7.5.1) or (7.5.2) still holds. The same argument used in the proof of Theorem 7.2.1 or 7.2.2 can be borrowed to prove I_2 and I_3 (or I_5 and I_6) asymptotically zero. For I_1 (or I_4) we have

$$I_1(x) = \frac{1}{\sqrt{n}} \sum_{j=1}^n (\varepsilon_j^2 - (\sigma^2 + s(x_j)/n^a))(I(x_j \le x) - F(x))$$
$$+ \frac{n^{1/2-a}}{n} \sum_{j=1}^n s(x_j)(I(x_j \le x) - F(x)).$$

The first sum converges weakly to B_1 and the second tends to infinity or SF in Theorem 7.2.3 corresponding to $0 \le a < 1/2$ or $a = 1/2$.   □

7.5.3 Proofs of the Theorems in Section 3

Proof of Theorem 7.3.1. Let us deal with T_{n1}^* in the linear model case first. Similar to T_n, T_{n1}^* can be decomposed as

$$T_{n1}^*(x) = \frac{1}{\sqrt{n}} \sum_{j=1}^n (\hat{\varepsilon}_j^*)^2 (I(x_j \le x) - F_n(x))$$
$$- \frac{2}{\sqrt{n}} (\beta^* - \hat{\beta}) \sum_{j=1}^n \varepsilon_j^*(x_j - \bar{x})(I(x_j \le x) - F_n(x))$$
$$+ \frac{1}{\sqrt{n}} (\beta^* - \hat{\beta})^2 \sum_{j=1}^n (x_j - \bar{x})^2 (I(x_j \le x) - F_n(x)) + O_p(1/\sqrt{n})$$
$$=: I_1^*(x) - I_2^*(x) + I_3^*(x) + O_p(1/\sqrt{n}). \tag{7.5.8}$$

From the process of drawing the bootstrap data, b_i^* are independent with mean zero when the original data are given. This means that I_2^* defined by (7.3.1) is a conditionally centered process. If the weak convergence of I_2^* can be verified, we then can bind $(\beta^* - \hat{\beta})I_2^*$ by $O_p(1/\sqrt{n})$ as $(\beta^* - \hat{\beta}) = O_p(1/\sqrt{n})$. It is easy to see that $(\beta^* - \hat{\beta})^2 I_3^*$ can be bounded by $O_p(1/\sqrt{n})$ where I_3^* is defined by (7.3.1). We now prove that for almost all sequences $\{(x_1, y_1), \cdots, (x_n, y_n), \cdots\}$ I_2^* converges weakly to a Gaussian process I_2 with the covariance function $\sigma^2 E((X-EX)^2(I(X \le x) - F(x)(I(X \le x_1) - F(x_1)))$ for all x, x_1.

First, it is easy to check that the covariance function of I_2^* converges to the above one. Along with the proof of Theorem VII 21 (Pollard (1984), p. 157-159) or mimicking that in Zhu (1993), all we need to do is to verify condition (16) in the Equicontinuity lemma (Pollard (1984), p.150). It can be done by noticing that the class of the functions $(X - EX)\hat{\varepsilon}^*(I(X \le x) - F_n(x))$ over all indices x is a VC class. Hence condition (16) holds true. We can verify that I_1^* is a conditionally centered process and converges weakly to B_1. Hence the proof for the linear model case is finished. In the non-parametric regression case, T_{n1}^* is analogous to I_1^* in the linear model case and similar arguments can be applied. □

Proof of Theorem 7.3.2. It is easy to see that, together with the convergence of $\hat{\beta}$ in the linear model case and of $\hat{\phi}$, the covariance function of $T_n(E_n, \cdot)$ converges to that of T_n and finite-dimensional convergence of $T_n(E_n, \cdot)$ also holds. Therefore, it suffices to show uniform tightness. Let $g_n(X, Y, t) = (\hat{\varepsilon}^2 - \sigma^2)(I(X \le t) - F_n(t))$. For given $\{(x_1, y_1) \cdots, (x_n, y_n)\}$, define $d_n(t, s) = \sqrt{P_n(g_n(X, Y, t) - g_n(X, Y, s))^2}$, the $L^2(P_n)$ seminorm, where P_n is the empirical measure based on $\{(x_1, y_1) \cdots, (x_n, y_n)\}$ and for any function of (X, Y), $P_n f(X, Y)$ denotes the average value of n values $f(X_1, Y_1), \ldots, f(X_n, Y_n)$. For uniform tightness, all we need to do is to prove that for any $\eta > 0$ and $\epsilon > 0$, there exists a $\delta > 0$ such that

$$\limsup_{n \to \infty} P\{\sup_{[\delta]} |T_n(E_n, t) - T_n(E_n, s)| > \eta | X_n, Y_n\} < \epsilon \qquad (7.5.9)$$

where $[\delta] = \{(t, s) : d_n(t, s) \le \delta\}$ and $(X_n, Y_n) = \{(x_1, y_1), \cdots, (x_n, y_n)\}$.

Since the limiting property with $n \to \infty$ is investigated n will be always considered to be large enough below simplify some arguments of the proof. Let $g(X, Y, t) = (\varepsilon^2 - \sigma^2)(I(X \le t) - F_n(t))$, $\mathcal{G} = \{g(\cdot, t) : t \in R^d\}$ and $d(t, s) = \sqrt{P_n(g(X, Y, t) - g(X, Y, s))^2}$. By the convergence of $\hat{\beta}$ in the linear model and of $\hat{\phi}$ in non-parametric model, we have that $\sup_{t,s} |d_n(t, s) - d(t, s)| \to 0$ in probability. Hence for large n

$$P\{\sup_{[\delta]} |T_n(E_n, t) - T_n(E_n, s)| > \eta | X_n, Y_n\}$$

$$\le P\{\sup_{<2\delta>} |T_n(E_n, t) - T_n(E_n, s)| > \eta | X_n, Y_n\} \qquad (7.5.10)$$

where $< 2\delta >= \{(t, s) : d(t, s) \leq 2\delta\}$.

In order to apply the chaining lemma (e.g. Pollard (1984), p.144), we need to check that

$$P\{|T_n(E_n, t) - T_n(E_n, s)| > \eta\, d(t, s)|X_n, Y_n\} < 2\exp(-\eta^2/2) \quad (7.5.11)$$

and

$$J_2(\delta, d, \mathcal{G}) = \int_0^\delta \{2\log\{(N_2(u, d, \mathcal{G}))^2/u\}\}^{1/2} du \quad (7.5.12)$$

is finite for small $\delta > 0$ where the covering number $N_2(u, d, \mathcal{G})$ is the smallest m for which there exist m points $t_1, \ldots, t_m$ such that $\min_{1 \leq i \leq m} d(t, t_i) \leq u$ for every $t \in A$, (7.5.11) can be derived by the Hoeffding inequality and (7.5.12) is implied by the fact that $\mathcal{G}$ is a VC class and $N_2(u, d, \mathcal{G}) \leq c\,u^w$ for some constants c and w. Invoking the chaining lemma, there exists a countable dense subset $< 2\delta >^*$ of $< 2\delta >$ such that, combining with $J_2(\delta, d, \mathcal{G}) \leq c\,u^{1/2}$ for some $c > 0$,

$$P\{ \sup_{<2\delta>^*} \sqrt{n}|T_n(E_n, t) - T_n(E_n, s)| > 26cd^{1/2}|X_n, Y_n\}$$
$$\leq 2c\delta. \quad (7.5.13)$$

The countable dense subset $< 2\delta >^*$ can be replaced by $< 2\delta >$ itself because $T_n(E_n, t) - T_n(E_n, s)$ is a right-continuous function w.r.t. t and s. Together with (7.5.10), the proof is concluded from choosing δ small enough. $\qquad\square$

8

Checking the Adequacy of a Varying-Coefficients Model

8.1 Introduction

The defining characteristic of a longitudinal study is that individuals are measured repeatedly through time and a prime objective of the analysis is to evaluate the change of the mean response over time and the effects of the explanatory variables on the mean response. Recently, many efforts have been made towards varying-coefficient models in longitudinal analysis because the existing parametric and nonparametric approaches may be either too restrictive to accommodate the unknown shapes of the curves or lacking the specific structures of being biologically interpretable for many situations, among others, Hoover, Rice, Wu and Yang (1998), Wu Chiang and Hoover (1998), Fan and Zhang (1999), Wu and Chiang (2000), Fan and Zhang (2000), Huang, Wu and Zhou (2002), Chiang, Rice and Wu (2001), Wu and Liang (2004) and Huang, Wu and Zhou(2004).

The motivating example of some relevant papers above is the Multicenter AIDS Cohort Study, the data include the repeated measurements of physical examinations, laboratory results and CD4 cell counts and percentages of 283 homosexual men who became HIV-positive between 1984 and 1991. Since CD4 cells are vital for immune function, as stated in Wu and Chiang (2000), CD4 cell counts and percentage, i.e., CD4 cell count divide by the total number of lymphocytes, are currently the most commonly used markers for the health status of HIV infected persons, it is important to build some statistical models for the CD4 cell counts or percentage. The data set is an important special case of longitudinal data since the covariate variables are independent of the time t, i.e. the observations are cross-sectional. For this special data set $\{(t_{ij}, Y_{ij}, X_i^T) : i = 1, \cdots, n; j = 1, \cdots, n_i\}$ where $X_i = (1, X_i^{(1)}, \cdots, X_i^{(k)})$ existed in several longitudinal studies. Wu and Chiang (2000) employed the following varying-coefficient model

$$Y(t) = X^\tau \beta(t) + \varepsilon(t) \qquad (8.1.1)$$

where $X = (1, X^{(1)}, \cdots, X^{(k)}, X^{(l)}, l = 1, \cdots, k$ are time independent co-variates, $\beta(t) = (\beta_0(t), \ldots, \beta_k(t))^\tau, \beta_r(t)$ are smooth functions of t taking values on the real line, $\varepsilon(t)$ is a mean zero stochastic process with variance $Var(\varepsilon(t)) = \sigma^2(t)$ and covariance $Cov(\varepsilon(t), \varepsilon(s)) = \rho_\varepsilon(t, s)$. Also $X^{(1)}, \ldots, X^{(k)}$ are assumed to be random.

In some relevant researches, the model checking for the covariate effects has received much attention through detecting whether certain coefficient functions in a varying-coefficient model are constant. Among others, Cai, Li and Fan (2000) suggested a goodness-of-fit test based on a non-parametric maximum likelihood ratio test to detect whether certain coefficient function in varying-coefficient model are constant or whether any covariates are statistically significant in the model. Fan and Zhang (2000) tested whether some covariate effects follow certain parametric forms; the test statistics are based on the maximum deviations of the estimated coefficient functions from the true coefficient functions. Huang, Wu and Zhou (2002) introduced a hypothesis testing procedure for testing constant covariate effects based on function approximations through basis expansions and resampling subject bootstrap. When $n_i = 1$ for all $i = 1, \cdots, n_i$ in (8.1.1), i.e., a special case of the models described by Hastie and Tibshirani (1993), Cai, Fan and Li (2000) suggested a goodness-of-fit test based on a nonparametric maximum likelihood ratio test to detect whether certain coefficient function in varying-coefficient model are constant or whether any covariates are statistically significant in the model; Fan and Zhang (2000) proposed a test to check whether some covariate effects follow certain parametric forms, the test statistics are based on the maximum deviations of the estimated coefficient functions from the estimated parametric coefficient functions. However, these studies are not for model (8.1.1) with a longitudinal data set. Some tests have been proposed to handle the testing problem for the model where X is the function of t. See Huang, Wu and Zhou (2002) and Sun and Wu (2004), Clearly, model (8.1.1) is a special case. However, the asymptotical properties of test statistics and this bootstrap procedure used to evaluate the null distribution in Huang, Wu and Zhou (2002) were not derived theoretically even in checking whether the varying-coefficients are constants.

On the other hand, note that it is a piece of useful information that X is not a function of t. Clearly, although the tests designed for those more general models could be applicable, the information provided by the data set was not fully used and then in our setup those tests would not be powerful.

Therefore, in our setup, it is of great interest to construct tests to check whether some covariate effects in (8.1.1) follow certain parametric forms, and also more simply, detect whether certain coefficient functions in a varying-coefficient model are constant. However, to our knowledge, no reference has investigated the hypothesis testing for model (8.1.1) with the longitudinal data set.

This chapter aims to develop a global test procedure to assess the adequacy of the above model. Most of the materials are from Xu and Zhu

(2004). We focus on our attention to longitudinal samples with a real time t, a time-dependent response variable $Y(t)$, and a time-independent covariate vector $X = (X^{(0)}, \ldots, X^{(k)})^\tau$ with real valued $X^{(l)}$. Like usual linear models with an intercept term, we set $X^{(0)} \equiv 1$ and denote by (Y_{ij}, X_i, t_{ij}) the jth measurement of $(Y(t), X, t)$ for the i of the n independent subjects, where $X_i = (X_i^{(0)}, \ldots, X_i^{(k)})^\tau, X_i^{(0)} \equiv 1$ and $j = 1, 2, \ldots, n_i$. Assume that X_i are i.i.d. with distribution function F and t_{ij} are i.i.d. with distribution function G.

For model (8.1.1), when $E(XX^\tau)$ is invertible, $\beta(.)$ is uniquely defined and satisfies:

$$\beta(t) = (E(XX^\tau))^{-1} E(XY(t)) \qquad (8.1.2)$$

As the popularly used conventional diagnostics, we want to test whether $\beta_r(\cdot) = 0$ for a component r of interest with $0 \le r \le k$. We consider a more general problems. For any r with $1 \le r \le k$, let $\beta_r(\Theta) = \{\beta_r(.) \equiv \beta_r(., \theta); \theta \in \Theta\}$ be a family of parametric functions on an open subspace Θ of R^d for some $d \ge 1$. When a particular $\beta_r(.)$ is of interest, the problem can be formulated as testing the null hypothesis:

$$H_0 : \beta_r(\cdot) = \beta_r(\cdot, \theta_0) \text{ for some } \theta_0 \in \Theta \qquad (8.1.3)$$

against the saturated alternative

$$H_1 : \beta_r(\cdot) \neq \beta_r(\cdot, \theta_0) \text{ for any } \theta_0 \in \Theta \qquad (8.1.4)$$

For model (8.1.1) with the typical longitudinal sample we adopt a globally smoothing method to construct test statistic. Since t is real-valued time, the innovation approach proposed by Stute, Thies and Zhu (1998) can be applied. Through this method, some optimal tests can be constructed. See Stute (1997) and Stute, Thies and Zhu (1998). Furthermore, we also consider the NMCT approximation to the sampling null distribution. Comparing with a bootstrap, the NMCT approximation is much less computational burden and is the proven powerful tool. See Zhu (2003) in the conventional regression settings. It is worthwhile to mention that the applications of these two approaches are not trivial at all because the model structure with longitudinal data is much more complex than the ordinary regression model structure and then the construction of test and the study of its properties need to be delicate.

8.2 Test Procedures

Let e_{rl} be the $(r+1, l+1)$th element of $(E(XX^\tau))^{-1}$, $Z_{ir} = \sum_{l=0}^{k}(e_{ri}X_i^l)$, $Z_{ijr} = Z_{ir}Y_{ij}$. It can be deduced from (8.1.2) that, for $r = 0, 1, \ldots, k$,

$$\beta_r(t) = E\{Z_{ijr}|t_{ij} = t\} \qquad (8.2.1)$$

Under H_0, that is, $\beta_r(.) = \beta_r(.,\theta_0)$ for some $\theta_0 \in \Theta$, we have $E\{Z_{ijr} - \beta_r(t_{ij},\theta_0)|t_{ij} = t\} = 0$, i.e. $E\{Z_{ijr}|t_{ij} = t\} = \beta_r(t,\theta_0)$, $(i = 1,2,\ldots,n; j = 1,2,\ldots,m_i)$. Now a new model can be written as:

$$Z_{ijr} = \beta_r(t_{ij},\theta_0) + \varepsilon_{ij}, \quad (i = 1,2,\ldots,n; j = 1,2,\ldots,m_i) \qquad (8.2.2)$$

In this new model, we assume $E(\varepsilon_{ij}|t_{ij}) = 0, Var(\varepsilon_{ij}|t_{ij}) = \sigma^2(t_{ij})$ and that for each i, $t_{ij}, i = 1,2,\ldots,n; j = 1,2,\ldots,m_i$ are i.i.d. random variables drawn from a continuous distribution F. Since $E\{Z_{ijr} - \beta_r(t_{ij},\theta_0)|t_{ij} = t\} = 0$ for all t is equivalent to $R(t) = E\{Z_{ijr} - \beta_r(t_{ij},\theta_0)I(t_{ij} \le t)\} = 0$, for all t, under H_0,

$$T = \int R(t)^2 dF(t) = 0 \qquad (8.2.3)$$

where F is the distribution of t_{ij}.

Let $N = \sum_{i=1}^n m_i$ and $R_N(t) = \frac{1}{\sqrt{N}} \sum_{i=1}^n \sum_{j=1}^{m_i} \{(Z_{ijr} - \beta_r(t_{ij},\theta_0))I(t_{ij} \le t)\}$ as an empirical version of $R(t)$ if both Z_{ijr} and $\beta_r(t_{ij},\theta_0)$ are known, otherwise it should be $\tilde{R}_N(t) = \frac{1}{\sqrt{N}} \sum_{i=1}^n \sum_{j=1}^{m_i} \{(\hat{Z}_{ijr} - \beta_r(t_{ij},\theta_N))I(t_{ij} \le t)\}$ where θ_N and $\hat{Z}_{ijr}$ are the estimators of θ_0 and Z_{ijr} respectively. The estimator θ_N of θ_0 is given by the least squares method, i.e.

$$\theta_N = \arg\min_\theta \frac{1}{N} \sum_{i=1}^n \sum_{j=1}^{m_i} (Z_{ijr} - \beta_r(t_{ij},\theta))^2$$

let $\hat{\Sigma} = \frac{1}{n} \sum_{i=1}^n X_i X_i^\tau$, $\hat{e}_{rl}$ be the $(r+1,l+1)$th element of $\hat{\Sigma}^{-1}$, $\hat{Z}_{ri} = \sum_{l=0}^k (\hat{e}_{ri} X_i^l)$, $\hat{Z}_{ijr} = \hat{Z}_{ri} Y_{ij}$.

Define a test statistic

$$T_N = \int \tilde{R}_N(t)^2 dF_N(t) = \frac{1}{N} \sum_{i=1}^n \sum_{j=1}^{m_i} \tilde{R}_N(t_{ij})^2 \qquad (8.2.4)$$

where F_N is the empirical distribution based on $\{t_{ij}; (i = 1,2,\ldots,n; j = 1,2,\ldots,m_i)\}$. The null hypothesis H_0 is rejected for large values of T_N.

8.3 The Limit Behavior of Test Statistic

We first cite a standard result of least squares estimator. See, Jennrich (1969).

Proposition 8.1. *Under H_0 and the regularity conditions:*
(i) A sequence of real valued responses Z_{ijr} has the structure

$$Z_{ijr} = \beta_r(t_{ij},\theta_0) + \varepsilon_{ij}, \quad (i = 1,2,\ldots,n; j = 1,2,\ldots,m_i)$$

where $\beta_r(\cdot, \cdot)$ are continuous functions, with given form, on a compact subset Θ of a Euclidean space and the ε_{ij} are i.i.d. errors with zero mean and finite variance $\sigma^2 > 0$, and θ_0 and σ^2 are unknown.

(ii) The tail cross product of $\beta_r(\theta) = \{\beta_r(t_{ij}, \theta) : i = 1, 2, \ldots, n; j = 1, 2, \ldots, m_i\}$ with itself exists and that $Q(\theta) = |\beta_r(\theta) - \beta_r(\theta_0)|^2$ has a unique minimum at $\theta = \theta_0$. The definition of tail cross product can be found in Jennrich (1969).

(iii) Put, for ; $l = 1, \ldots, d$,

$$\beta'_{rk}(\theta) = \{\frac{\partial \beta_r(t_{ij}, \theta)}{\partial \theta_k} : i = 1, \ldots, n; j = 1, \ldots, m_i\} : k = 1, \ldots, d$$

$$\beta'_{rkl}(\theta) = \{\frac{\partial^2 \beta_r(t_{ij}, \theta)}{\partial \theta_k \partial \theta_l} : i = 1, \ldots, n; j = 1, \ldots, m_i\} : k = 1, \ldots, d$$

every element of $\beta'_{rk}(\theta)$ and $\beta'_{rkl}(\theta)$ exists and is continuous on Θ and that all tail cross products of the form $[f, h]$, where $f, h = \beta_r, \beta'_{rk}(\theta), \beta'_{rkl}(\theta)$, exist.

(iv) The true parameter vector θ_0 is an interior point of Θ and the matrix $a(\theta_0)$ is non-singular where $a(\theta) = ((\frac{\partial \beta_r(., \theta)}{\partial \theta_i} \times \frac{\partial \beta_r(., \theta)}{\partial \theta_j})_{ij}$.

Then $N^{\frac{1}{2}}(\theta_N - \theta_0)$ converges in distribution to a centered normal distribution with covariance structure $\sigma^2 a^{-1}(\theta_0)$.

We now state some regularity conditions for the need of the process convergence:

(A) (i) $\beta_r(t_{ij}, \theta)$ is continuously differentiable with respect to θ in the interior set of Θ. Let

$$g(t_{ij}, \theta) = grad_\theta(\beta_r(t_{ij}, \theta)) = (g_1(t_{ij}, \theta), \ldots, g_d(t_{ij}, \theta))^\tau,$$

and assume that

(A) (ii) $|g_i(t_{ij}, \theta)| \leq M(t_{ij})$ for all $\theta \in \Theta$ and $1 \leq i \leq d$ for an F-integrable function M.

(A) (iii) $\sigma^{-1}(t_{ij})|g_i(t_{ij}, \theta)| \leq M(t_{ij})$ for all $\theta \in \Theta$ and $1 \leq i \leq d$.

Set: $G(x, \theta) = \int_{-\infty}^{x} g(u, \theta) F(du)$. Under the condition of $(A)(i)$ and (ii), we can get the following theorem:

Theorem 8.2. Under H_0 and condition A, $\tilde{R}_N = \{\tilde{R}_N(t) : t \in R^1\}$ converges in distribution to a process

$$\tilde{R}_\infty = B - G^\tau(x, \theta_0) N$$

in the Skorohod space $D[-\infty, +\infty]$, where B is a centered Brownian motion with covariance function $Cov(B(x_1), B(x_2)) = \psi(x_1 \wedge x_2)$, where $\psi(x) = \int_{-\infty}^{x} Var(Z_{ijr}|t_{ij} = t) F(dt), t_{ij} \sim F$ and N is a centered normal vector with covariance $\sigma^2 a^{-1}(\theta_0)$. The convergence of the process implies that T_N converges in distribution to $T = \int \tilde{R}_\infty(t)^2 dF(t)$.

It is clear that the distribution of T_N and of T are intractable. Therefore, for determining critical values, we have to consider an approximation or a test statistic based on $\tilde{R}_N$ but having a tractable distribution. In the following, we introduce two approaches.

8.3.1 Innovation Process Approach

This approach has been used in Stute, Thies and Zhu (1998). We now apply it to the situation with longitudinal data. First, introduce scale invariant versions of R_N and $\tilde{R}_N$, namely,

$$R_N^0(t) = N^{-1/2} \sum_{i=1}^{n} \sum_{j=1}^{m_i} I(t_{ij} \leq t)\sigma^{-1}(t_{ij})(Z_{ijr} - \beta_r(t_{ij}, \theta_0))$$

and

$$\tilde{R}_N^0(t) = N^{-1/2} \sum_{i=1}^{n} \sum_{j=1}^{m_i} I(t_{ij} \leq t)\sigma^{-1}(t_{ij})(\hat{Z}_{ijr} - \beta_r(t_{ij}, \theta_N))$$

Replacing condition $(A)(ii)$ by $(A)(iii)$, we can obtain the convergence of R_N^0 in distribution to the process B_0 where B_0 is a centered Brownian motion with covariance function $Cov(B_0(x_1), B_0(x_2)) = F(x_1 \wedge x_2)$, and $\tilde{R}_N^0$ converges in distribution to a process $B_0 - G_0(x, \theta_0)^\tau N_0$ where

$$G_0(x, \theta_0) = \int_{-\infty}^{x} \sigma^{-1}(t)g(t, \theta_0)F(dt)$$

and N_0 is a standard normal vector.

A detailed study of $B_0 - G_0(x, \theta_0)^\tau N_0$ is difficult. As a result, a strategy will be to first transform $B_0 - G_0(x, \theta_0)^\tau N_0$ into B_0. That is, we construct a linear transformation L satisfying $LB_0 = B_0$ in distribution and $L(G_0(x, \theta_0)^\tau N_0) \equiv 0$. Below, we present the form of L. Set

$$A(s) = \int_{s}^{+\infty} g(t, \theta_0)g(t, \theta_0)^\tau \sigma^{-2}(t)F(dt)$$

a positive definite $d \times d$-matrix, defining

$$(Lf)(s) = f(s) - \int_{-\infty}^{s} \sigma^{-1}(t)g(t, \theta_0)^\tau A^{-1}(t)[\int_{t}^{+\infty} \sigma^{-1}(z)g(z, \theta_0)f(dz)]F(dt)$$

With these L, it is easy to prove $L(G_0(s, \theta_0)^\tau N_0) \equiv 0$ and since L is a linear operator, LB_0 is a centered Gaussian process. To prove $LB_0 = B_0$, we only need to show that

$$Cov(LB_0(r), LB_0(s)) = Cov(B_0(r), B_0(s)) = F(r \wedge s). \qquad (8.3.1)$$

A proof of (8.3.1) will be deferred to Section 8.5.

Theorem 8.3. *Under condition A, the regularity conditions in Proposition 3.1 and the null hypothesis H_0 with the assumption that $A(x)$ is nonsingular for all x, we have that, in distribution,*

$$L(B_0 - G_0(x,\theta_0)^\tau N_0) = L(B_0) = B_0 \tag{8.3.2}$$

and $L\tilde{R}_N^0$ converges in distribution to B_0 in the Skorohod space $D[-\infty, +\infty]$.

From statistical applications such as goodness-of-fit testing, Theorem 8.3 is still inappropriate since both $\tilde{R}_N^0$ and L involve unknown quantities like $\sigma^2(t_{ij}), \theta_0$ and $F(t_{ij})$. To apply our method to a given set of data, the transformation L, for example, needs to be replaced by an empirical analog, L_N. We then need to show that the resulting processes have the same limit as $L\tilde{R}_N^0$.

In the homoscedastic case we simply have to replace $\tilde{R}_N^0$ by $\sigma_N^{-1}\tilde{R}_N$, where σ_N^2 is the sum of the squared residuals and similarly in L. In the general heteroscedastic case, however, it is the function $\sigma^2(t_{ij})$ rather than the constant σ^2 which needs to be estimated from the data. Because $\sigma^2(t_{ij}) = E\{Z_{ijr}^2|t_{ij} = t\} - \beta_r(t_{ij})^2$, any consistent non-parametric regression curve estimator may serve as an empirical substitute for the conditional second moment. Under H_0, $\beta_r(t_{ij})^2$ may be estimated by $\beta_r(t_{ij}, \theta_N)^2$. As it turns out, this procedure works in principle, under some restrictive smoothness assumptions on $\sigma^2(t_{ij})$. A workable approach is the following: split the whole sample $\{(t_{ij}, Z_{ijr}), (i = 1, 2, \ldots, n; j = 1, 2, \ldots, m_i)\}$ into two parts, say S_1, S_2. We assume that $S_1 = \{(t_{ij}, Z_{ijr}) : i = 1, \ldots, n_1; j = 1, \ldots, m_i\}$ and $S_2 = \{(t_{ij}, Z_{ijr}) : i = n_1 + 1, \ldots, n; j = 1, \ldots, m_i\}$, the sizes of S_1 and S_2 equal $N_1 = \sum_{i=1}^{n_1} m_i$ and $N_2 = \sum_{n_1+1}^n m_i$, assuming that both N_1 and N_2 converge to infinity as $N \to \infty$. Then we estimate $\sigma^2(t_{ij})$ from the first part, say by $\sigma_{N_1}^2(t_{ij})$, and let the process based on the second part. This leads to the two processes

$$R_N^1(t) = N_2^{-1/2} \sum_{i=n_1+1}^n \sum_{j=1}^{m_i} I(t_{ij} \le t)\sigma_{N_1}^{-1}(t_{ij})[Z_{ijr} - \beta_r(t_{ij}, \theta_0)],$$

$$\tilde{R}_N^1(t) = N_2^{-1/2} \sum_{i=n_1+1}^n \sum_{j=1}^{m_i} I(t_{ij} \le t)\sigma_{N_1}^{-1}(t_{ij})[\hat{Z}_{ijr} - \beta_r(t_{ij}, \theta_{N_1})].$$

Finally, the transformation L_N is defined by

$$(L_N f)(s)$$
$$= f(s) - \int_{-\infty}^s \sigma_{N_1}^{-1}(t)g(t,\theta_{N_1})^\tau A_{N_1}^{-1}(t)[\int_t^{+\infty} \sigma_{N_1}^{-1}(z)g(z,\theta_{N_1})f(dz)]F_{N_1}(dt).$$

Here F_{N_1} is the empirical d.f. of $\{Z_{ijr}, (t_{ij}, Z_{ijr}) \in S_2\}$, the estimator θ_{N_1} and $\hat{Z}_{ijr}$ are computed from $\{Z_{ijr}, (t_{ij}, Z_{ijr}) \in S_2\}$, and

$$A_{N_1}(s) = \int_s^{+\infty} g(t, \theta_{N_1}) g(t, \theta_{N_1})^\top \sigma_{N_1}^{-2}(t) F_1(dt).$$

To demonstrate the effect of splitting the data into two parts, note that conditionally on the first N_1 data, R_N^1 is a sum of independent centered processes with the covariance function

$$K_{N_1}(r, s) = \int_{-\infty}^{r \wedge s} \sigma^2(t)/\sigma_{N_1}^2(t) F(t).$$

We shall see that under appropriate conditions:

$$sup_{r,s} E|K_{N_1}(r, s) - F(r \wedge s)| \to 0 \qquad (8.3.3)$$

which together with the above-mentioned independence of summands yields

$$R_N^1(t) \to B_0 \quad \text{in distribution.}$$

For $\sigma_{N_1}(t)$ in L_N we recall that under no condition other than square-integrability of Z_{ijr} do there exist universally consistent estimators of $\sigma_N^2(t)$ satisfying:

$$E \int |\sigma_{N_1}^2(t) - \sigma_N^2(t)| F(dt) \to 0 \quad N_1 \to \infty. \qquad (8.3.4)$$

For the convergence of the covariance function K_{N_1}, we need to assume that $\sigma^2(t)$ is bounded away from zero, that is $\sigma^2(t) \geq a > 0$ for some a. For theoretical purposes we also want to guarantee that the $\sigma_{N_1}^2(t)$ are bounded away from zero.

Theorem 8.4. *Under the condition of Theorem 8.3 and (8.3.4) with $\sigma_{N_1}^2$ being a universally consistent estimator of σ^2 bounded away from zero. We have, under the null hypothesis H_0,*

$$L_N \tilde{R}_N^1 \to B_0 \quad \text{in distribution in the} \quad D[-\infty, +\infty] \qquad (8.3.5)$$

The convergence of the above process implies that $\tilde{T}_N := \int (L_N \tilde{R}_N^1(t))^2 dF_N(t)$ converges in distribution to $\int B_0(t)^2 d(t).$

8.3.2 A Non-parametric Monte Carlo Test

For a comparison with the innovation process approach, we consider NMCT for determining critical values in this section. Note that under the null hypothesis H_0 and the regularity conditions in Proposition 8.1, we have

$$N^{1/2}(\theta_N - \theta_0) = N^{-1/2} \sum_{i=1}^{n} \sum_{j=1}^{m_i} l(t_{ij}, Z_{ijr}, \theta_0) + o_p(1)$$

where l is a vector-valued function such that

$$E(l(t_{ij}, Z_{ijr}, \theta_0)) = 0, E\{l(t_{ij}, Z_{ijr}, \theta_0)l(t_{ij}, Z_{ijr}, \theta_0)'\} = \sigma^2 a^{-1}(\theta_0).$$

Let $J(Z_r, T, \theta_0, t) = I(T \le t)(Z_r - \beta_r(T, \theta_0)) + E(T \le t)l(T, Z_r, \theta_0)$. Then

$$\tilde{R}_N(t) = \frac{1}{\sqrt{N}} \sum_{i=1}^{n} \sum_{j=1}^{m_i} J(Z_{ijr}, t_{ij}, \theta_0, t) + o_p(1).$$

The algorithm is as follows.

Step 1. Generate independent identically distributed random variables $e_{ij}, i = 1, \ldots, n; j = 1, \ldots, m_i$, each having bounded support with mean zero and variance one. Let $E_N := \{(e_{ij}, i = 1, \ldots, n; j = 1, \ldots, m_i\}$ and define the conditional counterpart of R_N as

$$\tilde{R}_N(t, E_N) = \frac{1}{\sqrt{N}} \sum_{i=1}^{n} \sum_{j=1}^{m_i} e_{ij} J(\hat{Z}_{ijr}, t_{ij}, \theta_N, t). \qquad (8.3.6)$$

The resulting test statistic is

$$T_N(E_N) = \int \tilde{R}_N(t, E_N)^2 dF_N(t). \qquad (8.3.7)$$

Step 2. Generate k sets of E_N, say $E_N^{(i)}, i = 1, \ldots, k$ and get k values of $T_N(E_N)$, say $T_N(E_N)^{(i)}, i = 1, \ldots, k$.

Step 3. The p-value is estimated by $\hat{p} = k/(m+1)$ where k is the number of $T_N(E_N)^{(i)}$'s which are larger than or equal to $T_N(E_N)$. Reject H_0 when $p \le \alpha$ for a designated level α.

The following result states the consistency of the NMCT approximation.

Theorem 8.5. *Under the null hypothesis H_0, and the conditions in Theorem 8.2, we have that, for almost all sequences $\{(X_i, Y_{ij}, t_{ij}), i = 1, \ldots, n; j = 1, \ldots, m_i\}$, the conditional distribution of $T_N(E_N)$ converges to the limiting null distribution of T_N.*

8.4 Simulation Study and Application

8.4.1 Simulation Study

We consider parametric model of (8.1.1). The model has the coefficient curves $\beta_0 = t + at^2$ and $\beta_1 = 1$ and the covariate vector $X = (1, X)^\tau$ where X is a standard normal random variable.

We generated the $i.i.d.$ time points $\{t_{ij}\}_{1\leq i\leq n, 1\leq m_i}$ from the uniform distribution on $[0,1]$, and the random errors of ε_{ij} from the mean zero Gaussian process with covariance function:

If $\quad i_1 = i_2, \quad Cov(\varepsilon(t_{i_1,j_1}), \varepsilon(t_{i_2,j_2})) = \lambda\exp(-\lambda|t_{i_1,j_1} - t_{i_1,j_1}|); \text{otherwise} \quad 0.$

The responses $\{Y_{ij}\}_{1\leq i\leq n, 1\leq m_i}$ are obtained by substituting the corresponding time points, covariate vectors, random errors and coefficient curves into (8.1.1).

In our simulation study we let $\beta_0(\Theta) = \{\beta_0(.) \equiv \theta t\}$ be the family of linear functions. Hence $H_0 : \beta_0(t) \in \beta_0(\Theta)$ holds with $\theta_0 = 1$ and if and only if $a = 0$. Throughout the simulations, the nominal level is $\alpha = 0.05$ and various values for $\lambda = 0.5, 1, 2$ and $a = 0, 0.5, 1, 1.5, 2, 2.5$ are considered. The number of independent subjects and measurement are $n = 20, 50$ and $m_i \equiv 20, 50$. For each sample, the p-value is determined using 1000 replications of the Monte Carlo procedures.

Table 8.1. Empirical powers of test T_n with $\alpha = 0.05$: $\lambda = 1$; *Innovation method*

a	0.00	0.50	1.00	1.50	2.00	2.50
$n = 20, m = 20$	0.0220	0.0740	0.2480	0.5320	0.8180	0.9220
$n = 20, m = 50$	0.0240	0.1740	0.6000	0.8960	0.9780	0.9960

Table 8.2. Empirical powers of test T_n with $\alpha = 0.05$: $\lambda = 1$; *NMCT*

a	0.00	0.50	1.00	1.50	2.00	2.50
$n = 20, m = 20$	0.0400	0.0933	0.2833	0.4667	0.7033	0.9000
$n = 20, m = 50$	0.0200	0.1467	0.5333	0.8667	0.9800	1.0000

From Tables 8.1 and 8.2, we can see that the tests based the two methods have good power performance even when the sample size of n is so small. It clearly shows the sensitiveness of the tests to the alternatives. It also shows that when the sample size is large with $m = 50$, the power increases greatly. Comparing the power performance of the two methods, we can find that they are comparable. See Figure 8.1 for a clearer picture. Considering the computation burden issue, we may recommend the innovation approach in the examples we conducted.

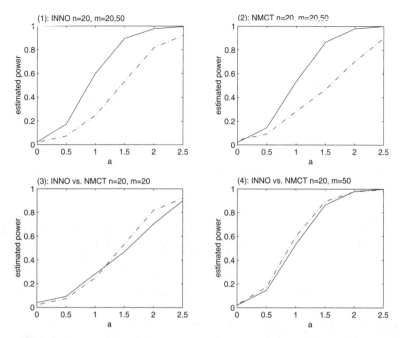

Fig. 8.1. (1) The dashdot line is the estimated power of Innovation method with $n = 20$ and $m = 20$ and the solid line is with $n = 20$ and $m = 50$; (2) The dashdot line is the estimated power of NMCT with $n = 20$ and $m = 20$ and the solid line is with $n = 20$ and $m = 50$; (3) The dashdot line is the estimated power of Innovation method with $n = 20$ and $m = 20$ and the solid line is for NMCT with $n = 20$ and $m = 20$; (4) The dashdot line is the estimated power of Innovation method with $n = 20$ and $m = 50$ and the solid line is for NMCT with $n = 20$ and $m = 50$.

8.4.2 Application to AIDS Data

As mentioned in the introduction, the data set considered here came from the Multi-Center AIDS Cohort Study. It contains the HIV status of 283 homosexual men who were infected with HIV during the follow-up period between 1984 and 1991. All individuals were scheduled to have their measurements made at semi-annual visits, but because many individuals missed some of their scheduled visits and the HIV infections happened randomly during the study, there are unequal numbers of repeated measurements and different measurement times per individual. Further details about the design, methods and medical implications of the study can be found in Kasolw et al.(1987).

The focus of our statistical analysis is to detect the effects of cigarette smoking, pre-HIV infection CD4 percentage and age at HIV infection on the mean CD4 percentage after the infection. Denote by t_{ij} the time in years of the jth measurements of the ith individual after HIV infection, by Y_{ij} the ith individual's CD4 at time t_{ij} and by $X_i^{(1)}$ the ith individual's smoking status; $X_i^{(1)}$ is 1 or 0 if the ith individual ever or never smoked cigarettes, respectively, after the HIV infection. For a clear biological interpretation of the coefficient functions , we define $X_i^{(2)}$ to be the ith individual's centered age at HIV infection, obtained by subtracting the sample average age at infection from the ith individual's age at infection. Similarly, $X_i^{(3)}$, the ith individual's centered pre-infection CD4 percentage, is computed by subtracting the average pre-infection CD4 percentage of the sample from the ith individual's actual pre-infection CD4 percentage. These covariates, except the time, are time-invariant.

The varying-coefficient model for Y_{ij}, t_{ij} and $X_i = (1, X_i^{(1)}, X_i^{(2)}, X_i^{(3)})$ is

$$Y_{ij} = \beta_0(t_{ij}) + X_i^{(1)}\beta_1(t_{ij}) + X_i^{(2)}\beta_2(t_{ij}) + X_i^{(3)}\beta_3(t_{ij}) + \epsilon_{ij}$$

where $\beta_0(t)$, the baseline CD4 percentage, represents the mean CD4 percentage t years after the infection for a non-smoker with average pre-infection CD4 percentage and average age at HIV infection, and the time-varying effects for cigarette smoking, age at HIV infection and pre-infection CD4 percentage on the post-infection CD4 percentage at time t are described by $\beta_1(t)$, $\beta_2(t)$ and $\beta_3(t)$ respectively.

In three prior analysis of the same dataset, Wu and Chiang (2000) and Fan and Zhang (2000) considered the nonparametric estimation of $\beta_l(t)(l = 0, 1, 2, 3)$ using locally smoothing methods. Huang, Wu and Zhou (2002) proposed globally smoothing methods to estimate $\beta_l(t)(l = 0, 1, 2, 3)$ and do some inference procedures to detect whether coefficient functions $\beta_l(t)(l = 0, 3)$ in a varying-coefficient model are constant or whether covariates $X_1^{(1)}, X_1^{(2)}$ are statistically significant in the model. The findings of Wu and Chiang (2000), Fan and Zhang (2000) and Huang, Wu and Zhou (2002) for estimating the varying-coefficient coefficients are basically the same, see Huang, Wu and Zhou (2002) for the estimation results.

We analysed the same data and introduced two methods to test whether some covariate effects follow certain parametric forms, we considered eight null hypotheses for coefficient curves, $\beta_i(t) = a_i + b_i t, i = 0, 1, 2, 3$ and $\beta_i(t) = c_i, i = 0, 1, 2, 3$, that is we first test whether the effect of smoking, age, baseline and Pre-CD4 is linear, then check whether they are constant effect. With 1000 replication times for p-values in NMCT method, the results are summarized in Table 8.3 at the 0.05 significance level. Since Huang, Wu and Zhou (2002) investigated the same data and considered four null hypotheses, $\beta_1(t) = 0$, $\beta_2(t) = 0$, $\beta_0(t) = c_0$ and $\beta_3(t) = c_3$, i.e. smoking has no effect, age has no effect, baseline effect is constant and pre-CD4 effect is constant respectively, their results are listed in Table 8.4 for compare.

Table 8.3. Application to AIDS data, $\alpha = 0.05$:

Null Hypothesis	Innovation p-value	NMCT p-value	Null Hypothesis	Innovation p-value	NMCT p-value
$\beta_0(t) = a_0 + b_0 t$	0.2490-0.2878	0.8900	$\beta_0(t) = c_0$	0.0000	0.0060
$\beta_1(t) = a_1 + b_1 t$	0.7012-0.8910	0.9880	$\beta_1(t) = c_1$	0.2490-0.2878	0.6000
$\beta_2(t) = a_2 + b_2 t$	0.7012-0.8910	0.9960	$\beta_2(t) = c_2$	0.0879-0.0999	0.4200
$\beta_3(t) = a_3 + b_3 t$	0.4672-0.5652	0.9760	$\beta_3(t) = c_3$	0.3346-0.2878	0.6660

Table 8.4. Result of Huang, Wu and Zhou(2002) $\alpha = 0.05$

Null hypothesis	p-value
$\beta_1(t) = 0$	0.176
$\beta_2(t) = 0$	0.301
$\beta_0(t) = c_0$	0.000
$\beta_3(t) = c_3$	0.059

From the results, all three methods have convincing evidence for rejecting the null hypothesis $\beta_0(t) = c_0$ and no sufficient evidence to reject other null hypothesis.

8.5 Proofs

As we described in Section 8.3, the following lemma can provide the basis for the result in Theorem 8.2.

Lemma 8.6. *We have*

$$Cov(LB_0(r), LB_0(s)) = Cov(B_0(r), B_0(s)).$$

Proof. Without loss of generality assume that $r \leq s$. According to the definition of L we have

$$(LB_0)(r) = B_0(r) - \int_{-\infty}^{r} \sigma^{-1}(t)g(t,\theta_0)^\tau A^{-1}(t)[\int_{t}^{+\infty} \sigma^{-1}(z)g(z,\theta_0)B_0(dz)]F(dt).$$

Note that the mean of LB_0 and of B_0 are zero. By some elementary calculation, we obtain that

$$
\begin{aligned}
&E(LB_0(r) \times LB_0(s)) \\
&= E\{B_0(r) \times B_0(s)\} \\
&\quad - E\{B_0(r) \times \int_{-\infty}^{s} \sigma^{-1}(t)g(t,\theta_0)^\tau A^{-1}(t)[\int_{t}^{+\infty} \sigma^{-1}(z)g(z,\theta_0)B_0(dz)]F(dt)\} \\
&\quad - E\{B_0(s) \times \int_{-\infty}^{r} \sigma^{-1}(t)g(t,\theta_0)^\tau A^{-1}(t)[\int_{t}^{+\infty} \sigma^{-1}(z)g(z,\theta_0)B_0(dz)]F(dt)\} \\
&\quad + E\{\int_{-\infty}^{r} \sigma^{-1}(t)g(t,\theta_0)^\tau A^{-1}(t)[\int_{t}^{+\infty} \sigma^{-1}(z)g(z,\theta_0)B_0(dz)]F(dt) \\
&\qquad \times \int_{-\infty}^{s} \sigma^{-1}(t)g(t,\theta_0)^\tau A^{-1}(t)[\int_{t}^{+\infty} \sigma^{-1}(z)g(z,\theta_0)B_0(dz)]F(dt)\}. \quad (8.5.1)
\end{aligned}
$$

Upon using rules for stochastic integrals, we can get that

$$E(LB_0(r) \times LB_0(s)) = Cov\{B_0(r), B_0(s)\}$$
$$- \int_{-\infty}^{s} \sigma^{-1}(t)g(t,\theta_0)^\tau A^{-1}(t)[\int_{t}^{r} \sigma^{-1}(z)g(z,\theta_0)F(dz)]F(dt)$$
$$- \int_{-\infty}^{r} \sigma^{-1}(t)g(t,\theta_0)^\tau A^{-1}(t)[\int_{t}^{s} \sigma^{-1}(z)g(z,\theta_0)F(dz)]F(dt)$$
$$- \int_{-\infty}^{r}\int_{-\infty}^{s} \sigma^{-1}(t)g(t,\theta_0)^\tau A^{-1}(t)A(t \vee z)A^{-1}(z)\sigma^{-1}(z)g(z)F(dz)F(dt).$$

Because the fourth element of the above equality is

$$\int_{-\infty}^{r}\int_{-\infty}^{s} \sigma^{-1}(t)g(t,\theta_0)^\tau A^{-1}(z)\sigma^{-1}(z)g(z)I(t \geq z)F(dz)F(dt)$$
$$+ \int_{-\infty}^{r}\int_{-\infty}^{s} \sigma^{-1}(t)g(t,\theta_0)^\tau A^{-1}(t)\sigma^{-1}(z)g(z)I(t < z)F(dz)F(dt)$$
$$= \int_{-\infty}^{r}\int_{-\infty}^{s} \sigma^{-1}(t)g(t,\theta_0)^\tau A^{-1}(z)\sigma^{-1}(z)g(z)I(t \geq z)F(dz)F(dt)$$
$$+ \int_{-\infty}^{r}\int_{-\infty}^{s} \sigma^{-1}(t)g(t,\theta_0)^\tau A^{-1}(t)\sigma^{-1}(z)g(z)I(t < z)F(dz)F(dt).$$

Finish the proof of the lemma by summation and application of the Fubini Theorem. $\qquad\square$

Lemma 8.7. *Under H_0, we have in probability and uniformly on compact set*

$$L\tilde{R}_N^0 = LR_N^0 + o_p(1). \tag{8.5.2}$$

Proof. By the definition of L,

$$(L\tilde{R}_N^0)(s) = \tilde{R}_N^0(s)$$
$$- \int_{-\infty}^{s} \sigma^{-1}(t)g(t,\theta_0)^\tau A^{-1}(t)[\int_{t}^{+\infty} \sigma^{-1}(z)g(z,\theta_0)\tilde{R}_N^0(dz)]F(dt),$$
$$(LR_N^0)(s) = R_N^0(s)$$
$$- \int_{-\infty}^{s} \sigma^{-1}(t)g(t,\theta_0)^\tau A^{-1}(t)[\int_{t}^{+\infty} \sigma^{-1}(z)g(z,\theta_0)R_N^0(dz)]F(dt).$$

Because

$$\int_{-\infty}^{s} \sigma^{-1}(t)g(t,\theta_0)^\tau A^{-1}(t)\left[\int_{t}^{+\infty} \sigma^{-1}(z)g(z,\theta_0)\tilde{R}_N^0(dz)\right]F(dt)$$
$$= N^{-1/2}\sum_{i=1}^{n}\sum_{j=1}^{m_j}\int_{-\infty}^{s} \sigma^{-1}(t)g(t,\theta_0)^\tau A^{-1}(t)I_{(t,\infty)}(t_{ij})\sigma^{-2}(t_{ij}) \times$$
$$g(t_{ij},\theta_0)(Z_{ijr} - \beta_r(t_{ij},\theta_0))F(dt)$$

and

$$\int_{-\infty}^{s} \sigma^{-1}(t)g(t,\theta_0)^{\tau}A^{-1}(t)[\int_{t}^{+\infty} \sigma^{-1}(z)g(z,\theta_0)R_N^0(dz)]F(dt)$$

$$= N^{-1/2}\sum_{i=1}^{n}\sum_{j=1}^{m_j}\int_{-\infty}^{s} \sigma^{-1}(t)g(t,\theta_0)^{\tau}A^{-1}(t)I_{(t,\infty)}(t_{ij})\sigma^{-2}(t_{ij}) \times$$

$$g(t_{ij},\theta_0)(\hat{Z}_{ijr} - \beta_r(t_{ij},\theta_N))F(dt).$$

Hence

$$(L\tilde{R}_N^0)(s) - (LR_N^0)(s)$$
$$= \{\tilde{R}_N^0(s) - R_N^0(s)\}$$

$$-\{N^{-1/2}\sum_{i=1}^{n}\sum_{j=1}^{m_j}\int_{-\infty}^{s} \sigma^{-1}(t)g(t,\theta_0)^{\tau}A^{-1}(t)I_{(t,\infty)}(t_{ij})\sigma^{-2}(t_{ij}) \times$$

$$g(t_{ij},\theta_0)(\hat{Z}_{ijr} - Z_{ijr})F(dt)\}$$

$$+\{N^{-1/2}\sum_{i=1}^{n}\sum_{j=1}^{m_j}\int_{-\infty}^{s} \sigma^{-1}(t)g(t,\theta_0)^{\tau}A^{-1}(t)I_{(t,\infty)}(t_{ij})\sigma^{-2}(t_{ij})g(t_{ij},\theta_0) \times$$

$$(\beta_r(t_{ij},\theta_N) - \beta_r(t_{ij},\theta_0))F(dt)\} \qquad (8.5.3)$$

We can get that, uniformly in $s \leq s_0$ for some fixed finite s_0, under H_0 and condition A

$$\tilde{R}_N^0 = R_N^0 - G_0(t_{ij},\theta_0)^{\tau}N^{1/2}(\theta_N - \theta_0) + o_p(1),$$

$$N^{-1/2}\sum_{i=1}^{n}\sum_{j=1}^{m_j}\int_{-\infty}^{s} \sigma^{-1}(t)g(t,\theta_0)^{\tau}A^{-1}(t)I_{(t,\infty)}(t_{ij})\sigma^{-2}(t_{ij})g(t_{ij},\theta_0)$$

$$\times(\beta_r(t_{ij},\theta_N) - \beta_r(t_{ij},\theta_0))F(dt)$$

$$= \int_{-\infty}^{s} \sigma^{-1}(t)g(t,\theta_0)^{\tau}A^{-1}(t)\int_{t}^{\infty} \sigma^{-2}(z)g(z,\theta_0)g(z,\theta_0)^{\tau}F(dz)F(dt) \times$$

$$N^{1/2}(\theta_N - \theta_0)$$

$$+o_p(1)$$
$$= G_0^{\tau}(s,\theta_0)N^{1/2}(\theta_N - \theta_0) + o_p(1),$$

and

$$\frac{1}{N^{1/2}}\sum_{i=1}^{n}\sum_{j=1}^{m_j}\int_{-\infty}^{s} \sigma^{-1}(t)g(t,\theta_0)^{\tau}A^{-1}(t) \times$$

$$I_{(t,\infty)}(t_{ij})\sigma^{-2}(t_{ij})g(t_{ij},\theta_0)(\hat{Z}_{ijr} - Z_{ijr})F(dt)$$

$$= o_p(1).$$

The proof is finished.   □

We are now in the position to prove the theorems.

Proof of Theorem 8.2.   Note that

$$\tilde{R}_N(t) = \frac{1}{\sqrt{N}} \sum_{i=1}^{n} \sum_{j=1}^{m} \{(\hat{Z}_{ijr} - \beta_r(t_{ij}, \theta_N))I(t_{ij} \le t)\}$$

$$= \frac{1}{\sqrt{N}} \sum_{i=1}^{n} \sum_{j=1}^{m} \{(Z_{ijr} - \beta_r(t_{ij}, \theta_0))I(t_{ij} \le t)\} - \frac{1}{\sqrt{N}} \sum_{i=1}^{n} \sum_{j=1}^{m} \{(\beta_r(t_{ij}, \theta_N)$$

$$- \beta_r(t_{ij}, \theta_0))I(t_{ij} \le t)\} + \frac{1}{\sqrt{N}} \sum_{i=1}^{n} \sum_{j=1}^{m} \{(\hat{Z}_{ijr} - Z_{ijr})I(t_{ij} \le t)\}.$$

It is not difficult to show that $\frac{1}{\sqrt{N}} \sum_{i=1}^{n} \sum_{j=1}^{m} \{(Z_{ijr} - \beta_r(t_{ij}, \theta_0))I(t_{ij} \le t)\}$ converges in distribution to a centered Brownian motion B with covariance function $Cov(B(x_1), B(x_2)) = \psi(x_1 \wedge x_2)$, $\frac{1}{\sqrt{N}} \sum_{i=1}^{n} \sum_{j=1}^{m} \{(\beta_r(t_{ij}, \theta_N) - \beta_r(t_{ij}, \theta_0))I(t_{ij} \le t)\}$ and $\frac{1}{\sqrt{N}} \sum_{i=1}^{n} \sum_{j=1}^{m} \{(\hat{Z}_{ijr} - Z_{ijr})I(t_{ij} \le t)\}$ converge, respectively, in distribution to $G^\tau(x, \theta_0)N$ and $o_p(1)$. The proof of this theorem is completed.   □

Proof of Theorem 8.3.   From Lemma 8.6, we can easily derive that, in distribution,

$$L(B_0 - G_0(x, \theta_0)^\tau N_0) = L(B_0) = B_0. \tag{8.5.4}$$

$LR_N^0 \to B_0$ and Lemma 8.7 implies that $L\tilde{R}_N^0 \to B_0$ in distribution. The proof of Theorem 8.3 is completed.   □

Proof of Theorem 8.4.   First, similar to the proof of Lemma 8.7, we obtain, upon using the fact that σ_{N_1} is bounded away from zero, that

$$L_N \tilde{R}_N^1 = L_N R_N^1 + o_p(1)$$

uniformly in $s \le s_0$. Second, we show that

$$L_N R_N^1 = L R_N^1 + o_p(1).$$

Note that

$$LR_N^1 - L_N R_N^1 + o_p(1)$$

$$= \int_{-\infty}^{s} \sigma_{N_1}^{-1}(t)g^\tau(t, \theta_{N_1})A_{N_1}^{-1}(t) \int_{t}^{\infty} \sigma_{N_1}^{-1}(z)g(z, \theta_{N_1})R_N^1(dz)F_{N_1}(dt)$$

$$- \int_{-\infty}^{s} \sigma_{N_1}^{-1}(t)g^\tau(t, \theta_0)A^{-1}(t) \int_{t}^{\infty} \sigma_{N_1}^{-1}(z)g(z, \theta_0)R_N^1(dz)F(dt)$$

$$= \int_{-\infty}^{s} \sigma_{N_1}^{-1}(t)g^{\tau}(t,\theta_0)A^{-1}(t) \int_{t}^{\infty} \sigma_{N_1}^{-1}(z)g(z,\theta_0)R_N^1(dz)\{F_{N_1}(dt) - F(dt)\}$$

$$+ \int_{-\infty}^{s} \{\sigma_{N_1}^{-1}(t)g^{\tau}(t,\theta_{N_1})A_{N_1}^{-1}(t) \int_{t}^{\infty} \sigma_{N_1}^{-1}(z)g(z,\theta_{N_1})R_N^1(dz)$$

$$- \sigma_{N_1}^{-1}(t)g^{\tau}(t,\theta_0)A^{-1}(t) \int_{t}^{\infty} \sigma_{N_1}^{-1}(z)g(z,\theta_0)R_N^1(dz)\}F_{N_1}(dt)$$

$$=: I_1 + I_2 \tag{8.5.5}$$

Putting

$$\alpha_N(t) = \sigma_{N_1}^{-1}(t)g^{\tau}(t,\theta_0)A^{-1}(t) \int_{t}^{\infty} \sigma_{N_1}^{-1}(z)g(z,\theta_0)R_N^1(dz)$$

it is not difficult to see that along with (8.3.3) and the boundedness of σ_{N_1}, the sequence α_N is tight. We can conclude that the first part I_1 of (8.5.5) tends to zero uniformly in $s \leq s_0$.

From condition (A) and the boundedness of σ_{N_1}, we obtain that the processes β_N defined by

$$\beta_N(s,\theta) = \int_{-\infty}^{s} \sigma_{N_1}^{-1}(t)g^{\tau}(t,\theta)A_{N_1}^{-1}(t,\theta) \int_{t}^{\infty} \sigma_{N_1}^{-1}(z)g(z,\theta)R_N^1(dz)F_{N_1}(dt)$$

are uniformly tight and continuous in θ. But $\theta_{N_1} \to \theta_0$ in probability so that the integral in the second part I_2 of (8.5.5) tends to zero in probability as N and $N_1 \to \infty$.

In the above L, σ has to be replaced by σ_{N_1}. Now mimic the proof of Theorem 8.3 to complete the proof of Theorem 8.4. $\qquad\square$

Proof of Theorem 8.5. By Wald's device, for almost all sequences $\{(X_i, Y_{ij}, t_{ij}) : i = 1, \ldots, n; j = 1, \ldots, m_i; N \to \infty\}$, all we need is to show that i) The covariance function of $R_N(., E_N)$ converges to that of R_N, ii) Finite distributional convergence of $R_N(., E_N)$ holds for any finite indices $\{t_{ij} : i = 1, \ldots, k; j = 1, \ldots, m_i\}$ and iii) Uniform tightness. The properties i) and ii) are easily verified, the details are omitted. Therefore, it suffices to show uniform tightness. It can be done along the lines of the proof of Theorem 3.2 of Zhu, Fujikoshi and Naito (2001). The detail is also omitted. $\qquad\square$

9

On the Mean Residual Life Regression Model

9.1 Introduction

The mean residual life (MRL) function $e(x)$ of a non-negative random variable X with finite mean is given by

$$e(x) = E(X - x | X > x) = S(x)^{-1} \int_x^T S(u)du, \qquad (9.1.1)$$

for $x < T$, where $S(x) = P(X > x)$ is the survival function and $T = \inf\{x : S(x) = 0\} \leq \infty$. Put $e(x) = 0$ whenever $x \geq T$. Like the hazard function, the MRL function completely determines the distribution via the inversion formula

$$S(x) = \frac{e(0)}{e(x)} \exp\left(- \int_0^x e(u)^{-1}du \right), \qquad (9.1.2)$$

for $x < T$, which is easily obtained from (9.1.1). This function turns out to be very useful in the study of replacement policies since the expected remaining life of a component gives us an indication of whether to replace an item or not. Other applications can be found in actuarial work, biomedical and demographic studies.

Oakes and Dasu (1990) proposed a new semiparametric proportional MRL model. Shortly thereafter, Maguluri and Zhang (1994) extended this model to a regression context. Their model specifies the conditional MRL function through

$$e(x|z) = \exp(-\beta'z)e_0(x), \qquad (9.1.3)$$

where $z' = (z_1, \ldots, z_p)$ is a p-dimensional covariate vector, $\beta' = (\beta_1, \ldots, \beta_p)$ is a vector of p regression parameters, and $e_0(x)$ is the MRL function corresponding to a baseline survival function S_0. They proposed two estimators for β. One is based on the underlying proportional hazards structure of the model while the other is called the simple estimator, based on the maximum likelihood equation of the exponential regression model. Both estimators were presented for uncensored data only.

The purpose of this chapter is to introduce a goodness-of-fit test for the MRL regression model and to examine two Monte Carlo approximations for the implementation of the test. Part of the material is from Zhu, Yuen and Tang (2002) and the proof of the theoretical results is from Zhu, Yuen and Tang (2000). In Section 9.2, we construct a test based on an appropriate stochastic process which is asymptotically Gaussian. Since the limiting null distribution of the test statistic is difficult to derive analytically, two Monte Carlo methods are applied to determine the p-values: the classical bootstrap and the NMCT approximation. It can be shown that both Monte Carlo methods yield asymptotically valid distributional approximations. These are described in Section 9.3. In Section 9.4, the performance of the two methods is assessed through a simulation study. The proofs of the main results are given in Section 9.5.

9.2 Asymptotic Properties of the Test Statistic

Suppose that $\{(X_1, Z_1), \ldots, (X_n, Z_n)\}$ are independent observations from a population having distribution function $F(x, z) = P(X \leq x, Z \leq z)$. The notation "$Z \leq z$" means that each component of Z is less than or equal to the corresponding component of z. After some algebraic manipulation, model (9.1.3) leads to

$$e_0(x) = G(x, z)^{-1} \int_x^T \int_0^z \exp(\beta' u)(t - x) F(dt, du), \qquad (9.2.1)$$

where $\int_0^z$ stands for $\int_0^{z_1} \cdots \int_0^{z_p}$ and $G(x, z) = F(T, z) - F(x, z)$. Let us denote the right-hand side of (9.2.1) by $A(x, z)$. Then our null hypothesis becomes

$$H_0 : \ e_0(x) = A(x, z) \ \text{ for all } \ z \ \text{ and all } \ x < T. \qquad (9.2.2)$$

In particular, when the model is true, the function A should be independent of z.

The so-called simple estimator $\hat{\beta}$ of β is defined as the solution of the equation

$$-\hat{U}(\beta) = n^{-1} \sum_{i=1}^{n} Z_i - \frac{\sum\limits_{i=1}^{n} X_i Z_i \exp(\beta' Z_i)}{\sum\limits_{i=1}^{n} X_i \exp(\beta' Z_i)} = 0. \qquad (9.2.3)$$

Let $t = (t_1, \cdots, t_p)'$ and $||t||$ be the L_2-norm of t. Assume that $E(\exp(t'Z)) < \infty$ with $||t|| < \epsilon$ and $\epsilon > 0$ and that $E((Z'Z + 1) \exp(2\beta'Z) X^2) < \infty$. Under these conditions, the arguments of Maguluri and Zhang (1994) derived the strong consistency and asymptotic normality of $\hat{\beta}$. Without further proof, the two conditions are assumed throughout the paper.

Our test statistic is based on the process, for any z_0,

$$V_n(x, z, z_0) = n^{\frac{1}{2}}(\hat{A}_n(x, z) - \hat{A}_n(x, z_0)), \qquad (9.2.4)$$

for all $x < T$ and $z > z_0$, where

$$\hat{A}_n(x, z) = \frac{\sum\limits_{i=1}^{n} \exp(\hat{\beta}' Z_i)(X_i - x)I(X_i > x, Z_i \leq z)}{\sum\limits_{i=1}^{n} I(X_i > x, Z_i \leq z)} \qquad (9.2.5)$$

is the empirical estimate of $A(x, z)$ in (9.2.2). Under the null hypothesis, the process V_n equals

$$V_n(x, z, z_0) = M_n(x, z) - M_n(x, z_0), \qquad (9.2.6)$$

where $M_n(x, z) = n^{\frac{1}{2}}(\hat{A}_n(x, z) - A(x, z))$. Our test of (9.1.3) will be based on the Cramér-von Mises type statistic

$$W_n = \int_0^b \int_{z_0}^b \int_0^T V_n^2(x, z, z_0) F_{Xn}(dx) F_{Zn}(dz) F_{Zn}(dz_0), \qquad (9.2.7)$$

where F_{Xn} and F_{Zn} are the empirical counterparts of the marginal distributions of X and Z, F_X and F_Z, respectively. The upper limit $b = (b_1, \ldots, b_p)$ in the first two integral signs is a vector of some arbitrarily large constants. One may simply choose b_j as the largest jth coordinate of Z in the sample. When the observed value of the test statistic is too large, the null hypothesis will be rejected.

Define

$$
\begin{aligned}
&f(x, z, z_0, X, Z; \beta, G) \\
&= \frac{(X - x)\exp(\beta'Z)I(X > x, z_0 < Z \leq z) - G(\beta, x, z, z_0)}{G(x, z)} \\
&\quad - \frac{(I(X > x, z_0 < Z \leq z) - G(x, z, z_0))G(\beta, x, z_0)}{G(x, z)G(x, z_0)} \\
&\quad + \left(Z' - \frac{XZ'\exp(\beta'Z)}{E(X\exp(\beta'Z))}\right)\Theta^{-1}(\beta)\left(\frac{\dot{G}(\beta, x, z, z_0)}{G(x, z)} - \frac{\dot{G}(\beta, x, z_0)G(x, z, z_0)}{G(x, z)G(x, z_0)}\right) \\
&\quad - \frac{((X - x)\exp(\beta'Z)I(X > x, Z \leq z_0) - G(\beta, x, z_0))G(x, z, z_0)}{G(x, z)G(x, z_0)} \\
&\quad + \frac{(I(X > x, Z \leq z_0) - G(x, z_0))G(x, z, z_0)G(\beta, x, z_0)}{G(x, z)G(x, z_0)^2}, \qquad (9.2.8)
\end{aligned}
$$

with $x < T$ and $z > z_0$ where

$$G(\beta, x, z) = \int_x^T \int_0^z \exp(\beta'u)(t - x)F(dt, du),$$

$$G(\beta, x, z, z_0) = G(\beta, x, z) - G(\beta, x, z_0),$$
$$G(x, z, z_0) = G(x, z) - G(x, z_0),$$
$$\Theta(\beta) = \left[\frac{\partial}{\partial \beta_1} U(\beta), \ldots, \frac{\partial}{\partial \beta_p} U(\beta) \right]_{p \times p}$$
$$U(\beta) = \frac{E(XZ \exp(\beta' Z))}{E(X \exp(\beta' Z))} - E(Z),$$
$$\dot{G}(\beta, x, z_0) = \left(\frac{\partial}{\partial \beta_1} G(\beta, x, z_0), \ldots, \frac{\partial}{\partial \beta_p} G(\beta, x, z_0) \right)',$$
$$\dot{G}(\beta, x, z, z_0) = \left(\frac{\partial}{\partial \beta_1} G(\beta, x, z, z_0), \ldots, \frac{\partial}{\partial \beta_p} G(\beta, x, z, z_0) \right)', \quad (9.2.9)$$

provided that $\Theta(\beta)$ is a $p \times p$ non-singular matrix. The following theorem states the distributional convergence of V_n.

THEOREM 9.2.1 *Under (9.2.2), V_n converges in the Skorohod space $D([0, T) \times [0, \infty]^{2p})$ in distribution to a centered Gaussian process V with covariance function, for any two pairs x, z, z_0 and x^0, z^0, z_0^0,*

$$Cov(f(x, z, z_0, X, Z; \beta, G), f(x^0, z^0, z_0^0, X, Z; \beta, G)). \quad (9.2.10)$$

Consequently, W_n of (9.2.7) converges in distribution to

$$W = \int_0^b \int_{z_0}^b \int_0^T V^2(x, z, z_0) F_X(dx) F_Z(dz) F_Z(dz_0). \quad (9.2.11)$$

Obviously, the complicated structure of the covariance function (9.2.10) does not allow for an analytic treatment of the involved distributions. To handle this problem, we use two resampling schemes to approximate the distribution of W.

We now study the power of our test statistic W_n for local alternatives converging to the null. Consider a sequence of $e(x|z)$ indexed by n

$$e_n(x|z) = e_0(x) \exp(-\beta' z - \delta(z) n^{-\frac{1}{2}}), \quad (9.2.12)$$

where $\delta(z)$ is an unknown function not depending on β. Parallel to (9.2.1), the baseline MRL function under (9.2.12) takes on the form

$$e_0(x) = G(x, z)^{-1} \int_x^T \int_0^z \exp(\beta' u + \delta(u) n^{-\frac{1}{2}})(t - x) F(dt, du).$$

From Taylor's expansion, it can be rewritten as

$$e_0(x) = G(x, z)^{-1} \int_x^T \int_0^z \exp(\beta' u)(1 + \delta(u) n^{-\frac{1}{2}} + o(n^{-\frac{1}{2}}))(t - x) F(dt, du).$$

Hence $\hat{A}_n$ of (9.2.5) under (9.2.12) becomes

$$
{}^{\delta}\hat{A}_n(x,z) = \frac{\sum\limits_{i=1}^{n} \exp(\hat{\beta}'Z_i)(X_i - x)I(X_i > x, Z_i \leq z)}{\sum\limits_{i=1}^{n} I(X_i > x, Z_i \leq z)}
$$

$$
+ \frac{n^{-\frac{1}{2}} \sum\limits_{i=1}^{n} \exp(\hat{\beta}'Z_i)(X_i - x)\delta(Z_i)I(X_i > x, Z_i \leq z)}{\sum\limits_{i=1}^{n} I(X_i > x, Z_i \leq z)} + o_p(n^{-\frac{1}{2}})
$$

$$
= \frac{\sum\limits_{i=1}^{n} \exp(\hat{\beta}'Z_i)(X_i - x)I(X_i > x, Z_i \leq z)}{\sum\limits_{i=1}^{n} I(X_i > x, Z_i \leq z)} + \frac{n^{-\frac{1}{2}}Q(x,z)}{G(x,z)} + o_p(n^{-\frac{1}{2}}),
$$

where $Q(x,z) = E(\exp(\beta'Z)(X-x)\delta(Z)I(X > x, Z \leq z))$. Therefore, parallel to V_n, under (9.2.12) we consider the process

$$
{}^{\delta}V_n(x,z,z_0) = n^{\frac{1}{2}}({}^{\delta}\hat{A}_n(x,z) - {}^{\delta}\hat{A}_n(x,z_0))
$$

$$
= n^{\frac{1}{2}}(\hat{A}_n(x,z) - \hat{A}_n(x,z_0)) + \left(\frac{Q(x,z)}{G(x,z)} - \frac{Q(x,z_0)}{G(x,z_0)} \right)
$$

$$
+ o_p(1). \tag{9.2.13}
$$

Note that the first term is the same as V_n of (9.2.4) and the second term is a non-random function. In other words, under alternative (9.2.12), the process V_n has a non-random shift.

9.3 Monte Carlo Approximations

A sensible way to cope with the intractability of the asymptotic null distribution of W_n is to seek Monte Carlo schemes to approximate critical values of the test. Here we study distributional approximations through the classical bootstrap and the NMCT method.

1. The Bootstrap Approximation.   Let $\{(X_1^*, Z_1^*), \ldots, (X_n^*, Z_n^*)\}$ be a bootstrap sample. Denote the bootstrap estimate of β by β^* as the solution of $-U^*(\beta) = 0$, where $U^*(\beta)$ is the bootstrap counterpart of (9.2.3). The weak convergence of β^* are discussed in Section 9.5. The bootstrap version of the process (9.2.6) is denoted by

$$
V_n^*(x,z,z_0) = M_n^*(x,z) - M_n^*(x,z_0), \tag{9.3.1}
$$

where

$$
M_n^*(x,z) = n^{\frac{1}{2}}(A_n^*(x,z) - \hat{A}_n(x,z)) \tag{9.3.2}
$$

and A_n^* is the bootstrap analogue of (9.2.5). Clearly, W_n^* takes on the form

$$W_n^* = \int_0^b \int_{z_0}^b \int_0^T (V_n^*(x, z, z_0))^2 F_{Xn}^*(dx) F_{Zn}^*(dz) F_{Zn}^*(dz_0), \qquad (9.3.3)$$

where F_{Xn}^* and F_{Zn}^* are the bootstrap marginal distributions. Following the work of Burke and Yuen (1995), one can show that (9.2.7) and (9.3.3) have the same limiting distribution under (9.2.2).

THEOREM 9.3.1 *Under (9.2.2), for a given sample $\{(X_i, Z_i); i = 1, \ldots, n\}$, W_n^* of (9.3.3) converges in distribution to W of (9.2.11) for almost all sequences $\{(X_1, Z_1), \ldots, (X_n, Z_n), \ldots\}$.*

2. The NMCT Approximation.   We first generate independent bounded random variables e_i, $i = 1, \ldots, n$, with zero mean and unit variance. Then, in this case, the NMCT process is defined as

$$V_n^R(x, z, z_0) = n^{-\frac{1}{2}} \sum_{i=1}^n e_i f(x, z, z_0, X_i, Z_i; \hat{\beta}, G_n),$$

where f is given in (9.2.8) and G_n is the empirical estimate of G. The resulting Cramér-von Mises statistic becomes

$$W_n^R = \int_0^b \int_{z_0}^b \int_0^T (V_n^R(x, z, z_0))^2 F_{Xn}(dx) F_{Zn}(dz) F_{Zn}(dz_0). \qquad (9.3.4)$$

Denote the observed value of W_n be $W_n^{R_0}$. Repeating the Monte Carlo procedure K times, we obtain $W_n^{R_1}, \ldots, W_n^{R_K}$. Since we reject the null hypothesis when W_n is too large, the estimated p-value is

$$\hat{p} = \frac{k}{K+1},$$

where k is the number of values in $\{W_n^{R_0}, \ldots, W_n^{R_K}\}$ that are larger than or equal to $W_n^{R_0}$. Therefore, for a given nominal level α, the null hypothesis is rejected whenever $\hat{p} \leq \alpha$.

THEOREM 9.3.2 *For almost all sequences $\{(X_1, Z_1), \ldots, (X_n, Z_n), \ldots\}$, we have, under (9.2.2), the conditional distribution of W_n^R given the data is asymptotically equal to the limiting distribution of W_n.*

Theorems 9.3.1 and 9.3.2 imply that both Monte Carlo approximations yield valid distributional approximations. It can be easily seen that asymptotically the test statistic has power one for fixed alternatives.

For local alternative (9.2.12), according to (9.2.13), we have the following result analogous to Theorem 9.3.2 based on the NMCT statistic (9.3.4).

With the NMCT, the distribution of W_n^R depends on

$$
{}^\delta V_n^R(x, z, z_0)
$$
$$
= n^{-1/2} \sum_{i=1}^n e_i \left(f(x, z, z_0, X_i, Z_i; \hat{\beta}, G_n) - n^{-\frac{1}{2}} \left(\frac{Q(x, z)}{G(x, z)} - \frac{Q(x, z_0)}{G(x, z_0)} \right) \right),
$$

if the data come from (9.2.12). This form of ${}^\delta V_n^R(x, z, z_0)$ implies that the distribution of W_n^R under (9.2.12) is asymptotically the same as that under (9.2.2), and hence that (9.2.2) and (9.2.12) yield the same critical value of the test asymptotically. The result is below.

THEOREM 9.3.3 *Under the alternative (9.2.12), the conditional distribution of W_n^R of (9.3.4) converges to that of W of (9.2.11) for almost all sequences* $\{(X_1, Z_1), \ldots, (X_n, Z_n), \cdots\}$.

9.4 Simulations

To demonstrate the performance of the two methods, a simulation study is carried through with sample sizes of $n = 50, 100, 200,$ and 300. We consider the case of a single covariate. The value of the covariate Z is either 0 or 1, with equal probability 0.5, so that we are in a two-sample case. From model (9.1.3), it is easy to see that the MRL of one sample is proportional to that of the other one with a factor $\exp(-\beta)$. Furthermore, (9.1.3) together with (9.1.1) and (9.1.2) yield

$$
S(x|z) = S_0(x) \left(\int_x^T \mu_0^{-1} S_0(u) du \right)^{\exp(\beta z) - 1}, \tag{9.4.1}
$$

where $\mu_0 = E(X) = e_0(0)$. Given a value of Z and a specific S_0, X can be generated from (9.4.1). In each simulation, 1000 independent bootstrap samples are drawn to compute the critical values for significance levels $\alpha = 0.1, 0.05$ and 0.025. Similarly, 1000 sets of $(e_1, \ldots, e_n)$, taking values of ± 1 with probability 0.5 were generated when performing the NMCT test.

Throughout, we assume that $S_0(x) = (1 - 0.5x)_+$ which comes from the class of survival distributions having a linear MRL function, introduced by Hall and Wellner (1984). The value of β used equals 0.8. The key results are summarized in Table 9.1. Each entry presents the proportion of cases when the null hypothesis was rejected, based on 1000 simulations. The bootstrap test looks rather conservative for small samples. As n increases, the actual levels converge to the true nominal levels from below. On the whole, the NMCT method produces values closer to the true levels.

Table 9.1. Empirical sizes for the resampling methods

	Bootstrap			NMCT		
n	$\alpha = 0.1$	$\alpha = 0.05$	$\alpha = 0.025$	$\alpha = 0.1$	$\alpha = 0.05$	$\alpha = 0.025$
50	0.077	0.031	0.010	0.104	0.061	0.031
100	0.073	0.036	0.020	0.071	0.039	0.021
200	0.091	0.037	0.018	0.096	0.049	0.022
300	0.100	0.047	0.024	0.106	0.052	0.033

As to empirical powers, we generate X from the proportional hazards model with the baseline distribution being Weibull. That is, the baseline is $\lambda(x|z) = \lambda_0(x)\exp(0.8z)$. The baseline hazard takes on the form $\lambda_0(x) = abx^{a-1}$. The two parameters a and b are set to be 3 and 0.00208, respectively, so that we have an increasing hazard function. Table 9.2 presents the empirical powers of the test for detecting alternatives to the proportional MRL assumption in the two-sample case. As shown in Table 9.2, the NMCT approximation outperforms the bootstrap. In particular, it works much better for small samples. Table 9.2 also indicates that for both methods, the power of the test approaches one as the sample size gets larger.

Table 9.2. Empirical powers for detecting non-proportional MRL

	Bootstrap			NMCT		
n	$\alpha = 0.1$	$\alpha = 0.05$	$\alpha = 0.025$	$\alpha = 0.1$	$\alpha = 0.05$	$\alpha = 0.025$
50	0.401	0.246	0.109	0.588	0.548	0.496
100	0.684	0.548	0.406	0.914	0.874	0.821
200	0.923	0.876	0.802	0.988	0.974	0.964
300	0.991	0.985	0.960	0.999	0.997	0.995

Generally speaking, both Monte Carlo approximations perform fairly well for moderate sample sizes. The above simulations show that the NMCT approximation is better for small sample sizes. To improve the performance of the naive bootstrap, one may use the bootstrap t method which requires a larger amount of computation. On the other hand, the NMCT procedure is computationally more efficient.

9.5 Proofs

In Section 9.2, four G functions, namely $G(x,z)$, $G(\beta,x,z)$, $G(x,z,z_0)$, and $G(\beta,x,z,z_0)$ were defined. Recall that the G_n's are their empirical counterparts with $G_n(x, z, z_0) = G_n(x, z) - G_n(x, z_0)$ and $G_n(\beta, x, z, z_0) = G_n(\beta, x, z) - G_n(\beta, x, z_0)$.

Proof of Theorem 9.2.1. By definition, the function $\hat{A}_n$ of (9.2.4) can be written as

$$\hat{A}_n(x, z) = \frac{G_n(\hat{\beta}, x, z)}{G_n(x, z)}.$$

For $x < T$ and $z > z_0$,

$$\hat{A}_n(x, z) - \hat{A}_n(x, z_0)$$
$$= \frac{G_n(\hat{\beta}, x, z)}{G_n(x, z)} - \frac{G_n(\hat{\beta}, x, z_0)}{G_n(x, z_0)}$$
$$= \frac{G_n(\hat{\beta}, x, z, z_0) - G(\hat{\beta}, x, z, z_0)}{G_n(x, z)} - \frac{(G_n(x, z, z_0) - G(x, z, z_0))G_n(\hat{\beta}, x, z_0)}{G_n(x, z)G_n(x, z_0)}$$
$$+ \left(\frac{G(\hat{\beta}, x, z, z_0)}{G_n(x, z)} - \frac{G(x, z, z_0)G_n(\hat{\beta}, x, z_0)}{G_n(x, z)G_n(x, z_0)} \right)$$
$$=: I_{n_1} - I_{n_2} + I_{n_3}.$$

From the theory of empirical processes, we can easily derive that $n^{\frac{1}{2}}(G_n(x, z) - G(x, z))$ and $n^{\frac{1}{2}}(G_n(\beta, x, z) - G(\beta, x, z))$ are asymptotically Gaussian. Moreover $G(\beta, x, z)$ is continuous with bounded first derivative with respect to β. Using these facts together with the strong consistency and asymptotic normality of $\hat{\beta}$, one can apply Taylor's expansion to obtain

$$n^{\frac{1}{2}}I_{n_1} = \frac{n^{\frac{1}{2}}(G_n(\beta, x, z, z_0) - G(\beta, x, z, z_0))}{G(x, z)} + o_p(1), \qquad (9.5.1)$$

$$n^{\frac{1}{2}}I_{n_2} = \frac{n^{\frac{1}{2}}(G_n(x, z, z_0) - G(x, z, z_0))G(\beta, x, z_0)}{G(x, z)G(x, z_0)} + o_p(1). \qquad (9.5.2)$$

Under H_0, $G(\beta, x, z, z_0) = G(x, z, z_0)G(\beta, x, z_0)/G(x, z_0)$. Therefore we have

$$I_{n_3} = \frac{1}{G_n(x, z)} \left(G(\hat{\beta}, x, z, z_0) - G(\beta, x, z, z_0) \right.$$
$$\left. - G(x, z, z_0) \left(\frac{G_n(\hat{\beta}, x, z_0)}{G_n(x, z_0)} - \frac{G(\beta, x, z_0)}{G(x, z_0)} \right) \right)$$
$$=: \frac{1}{G_n(x, z)} (I_{n_{31}} - G(x, z, z_0)I_{n_{32}}). \qquad (9.5.3)$$

Furthermore,

$$\frac{n^{\frac{1}{2}} I_{n_{31}}}{G_n(x,z)} = \frac{n^{\frac{1}{2}} (\hat{\beta} - \beta)' \dot{G}(\beta, x, z, z_0)}{G(x,z)} + o_p(1), \tag{9.5.4}$$

$$\frac{n^{\frac{1}{2}} G(x, z, z_0) I_{n_{32}}}{G_n(x,z)} = \frac{n^{\frac{1}{2}} (G_n(\hat{\beta}, x, z_0) - G(\hat{\beta}, x, z_0)) G(x, z, z_0)}{G_n(x,z) G_n(x, z_0)}$$

$$+ \frac{n^{\frac{1}{2}} (G(\hat{\beta}, x, z_0) - G(\beta, x, z_0)) G(x, z, z_0)}{G_n(x,z) G_n(x, z_0)}$$

$$- \frac{n^{\frac{1}{2}} (G_n(x, z_0) - G(x, z_0)) G(x, z, z_0) G(\beta, x, z_0)}{G_n(x,z) G(x, z_0) G_n(x, z_0)}$$

$$= \frac{n^{\frac{1}{2}} (G_n(\beta, x, z_0) - G(\beta, x, z_0)) G(x, z, z_0)}{G(x,z) G(x, z_0)}$$

$$- \frac{n^{\frac{1}{2}} (G_n(x, z_0) - G(x, z_0)) G(x, z, z_0) G(\beta, x, z_0)}{G(x,z) G(x, z_0)^2}$$

$$- \frac{n^{\frac{1}{2}} (\hat{\beta} - \beta)' \dot{G}(\beta, x, z_0) G(x, z, z_0)}{G(x,z) G(x, z_0)} + o_p(1), \tag{9.5.5}$$

where $\dot{G}(\beta, x, z_0)$ and $\dot{G}(\beta, x, z, z_0)$ are defined in (9.2.9). Note that $(\hat{\beta} - \beta) = -\Theta^{-1}(\beta)\hat{U}(\beta) + o_p(1/\sqrt{n})$. Combining (9.5.1)-(9.5.5), we have

$$V_n(x, z, z_0) = n^{\frac{1}{2}} (I_{n_1} - I_{n_2} + I_{n_3})$$

$$= n^{\frac{1}{2}} \left(\frac{G_n(\beta, x, z, z_0) - G(\beta, x, z, z_0)}{G(x,z)} \right.$$

$$- \frac{(G_n(x, z, z_0) - G(x, z, z_0)) G(\beta, x, z_0)}{G(x,z) G(x, z_0)}$$

$$- \frac{\hat{U}(\beta)' \Theta^{-1}(\beta)}{G(x,z)} \left(\dot{G}(\beta, x, z, z_0) - \frac{\dot{G}(\beta, x, z_0) G(x, z, z_0)}{G(x, z_0)} \right)$$

$$- \frac{(G_n(\beta, x, z_0) - G(\beta, x, z_0)) G(x, z, z_0)}{G(x,z) G(x, z_0)}$$

$$\left. + \frac{(G_n(x, z_0) - G(x, z_0)) G(x, z, z_0) G(\beta, x, z_0)}{G(x,z) G(x, z_0)^2} \right) + o_p(1)$$

$$= n^{-\frac{1}{2}} \sum_{j=1}^{n} f(x, z, z_0, X_j, Z_j; \beta, G) + o_p(1), \tag{9.5.6}$$

where $\hat{U}(\beta)$, $\Theta(\beta)$, and f are given in (9.2.3), (9.2.9) and (9.2.8), respectively.

Let the family of functions $\mathcal{F}$ be $\{f(x, z, z_0, X, Z; \beta, G); (x, z, z_0) \in [0, T) \times [0, \infty)^2, z_0 < z\}$. Since the functions $G(x, z)$ and $G(\beta, x, z)$ are absolutely continuous and the family of indicator functions $\{I(X > x, Z \leq z); (x, z) \in [0, T) \times [0, \infty)\}$ is a VC class, $\mathcal{F}$ is also a VC class (see Pollard (1984)). Lemma VII 15 and Theorem VII 21 of Pollard (1984, pages 150 and 157) imply that V_n converges in distribution to a centered Gaussian process with covariance function (9.2.10). The proof is complete.   □

For the proofs of Theorems 9.3.1, 9.3.2, and 9.3.3 to follow, we recall that all arguments hold for almost all sequences $\{(X_1, Z_1), \ldots, (X_n, Z_n), \ldots\}$.

Proof of Theorem 9.3.1. From Taylor's expansion , we have

$$0 = n^{\frac{1}{2}}\hat{U}(\hat{\beta}) = n^{\frac{1}{2}}\hat{U}(\beta) + \hat{\Theta}(\beta)n^{\frac{1}{2}}(\hat{\beta} - \beta) + o_p(1), \tag{9.5.7}$$

$$0 = n^{\frac{1}{2}}U^*(\beta^*) = n^{\frac{1}{2}}U^*(\hat{\beta}) + \Theta^*(\hat{\beta})n^{\frac{1}{2}}(\beta^* - \hat{\beta}) + o_p(1), \tag{9.5.8}$$

where $\hat{\Theta}$ and Θ^* are the empirical and bootstrap counterparts of Θ defined in (9.2.9), respectively. Following Burke and Yuen (1995), we obtain the convergence of $\hat{\Theta}(\beta)$ and $\Theta^*(\hat{\beta})$. Note that both converges to $\Theta(\beta)$. Then, applying techniques of Yuen and Burke (1997), one can show that $n^{\frac{1}{2}}(U^*(\hat{\beta}) - \hat{U}(\hat{\beta}))$ and $n^{\frac{1}{2}}(\hat{U}(\beta) - E(\hat{U}(\beta)))$ have the same normal distribution in the limit. The idea is to write $n^{\frac{1}{2}}(\hat{U}(\beta) - E(\hat{U}(\beta)))$ in terms of stochastic integrals involving empirical processes and β. Similar integral expressions can be written for $n^{\frac{1}{2}}(U^*(\hat{\beta}) - \hat{U}(\hat{\beta}))$ with the bootstrap empirical processes and $\hat{\beta}$. The theory of empirical processes and its bootstrap analogue leads to the desired result. By definition, $\hat{U}(\hat{\beta}) = E(\hat{U}(\beta)) = 0$. Therefore, it can be seen from (9.5.7) and (9.5.8) that $n^{\frac{1}{2}}(\beta^* - \hat{\beta})$ and $n^{\frac{1}{2}}(\hat{\beta} - \beta)$ have the same asymptotic distribution.

Denote the bootstrap version of the G's by G_n^*'s. From the standard bootstrap theory, all the G_n^*'s are consistent estimators of G's. Repeating the first step in the previous proof, we have

$$\begin{aligned}
&A_n^*(x, z) - A_n^*(x, z_0) \\
&= \frac{G_n^*(\beta^*, x, z)}{G_n^*(x, z)} - \frac{G_n^*(\beta^*, x, z_0)}{G_n^*(x, z_0)} \\
&= \frac{G_n^*(\beta^*, x, z, z_0) - G_n(\beta^*, x, z, z_0)}{G_n^*(x, z)} \\
&\quad - \frac{(G_n^*(x, z, z_0) - G_n(x, z, z_0))G_n^*(\beta^*, x, z_0)}{G_n^*(x, z)G_n^*(x, z_0)} \\
&\quad + \left(\frac{G_n(\beta^*, x, z, z_0)}{G_n^*(x, z)} - \frac{G_n(x, z, z_0)G_n^*(\beta^*, x, z_0)}{G_n^*(x, z)G_n^*(x, z_0)} \right) \\
&=: I_{n_1}^* - I_{n_2}^* + \tilde{I}_{n_3}^*.
\end{aligned} \tag{9.5.9}$$

Denote the first derivatives of $G_n^*(\beta, x, z, z_0)$ and $G_n(\beta, x, z, z_0)$ with respect to β by $\dot{G}_n^*(\beta, x, z, z_0)$ and $\dot{G}_n(\beta, x, z, z_0)$, respectively. It is easily seen that both $\dot{G}_n^*(\beta, x, z, z_0)$ and $\dot{G}_n(\beta, x, z, z_0)$ converge weakly to $\dot{G}(\beta, x, z, z_0)$. Taylor's expansion and the asymptotic normality of β^* yield

$$n^{\frac{1}{2}}I_{n_1}^* = \frac{n^{\frac{1}{2}}(G_n^*(\beta, x, z, z_0) - G_n(\beta, x, z, z_0))}{G(x, z)} + o_p(1).$$

Similarly, from the consistency of β^* and the G_n^* functions, we have

$$n^{\frac{1}{2}} I_{n_2}^* = \frac{n^{\frac{1}{2}}(G_n^*(x,z,z_0) - G_n(x,z,z_0))G(\beta,x,z_0)}{G(x,z)G(x,z_0)} + o_p(1).$$

Applying Theorem 2.4 of Giné and Zinn (1990), we immediately see from (9.5.1) and (9.5.2), that $n^{\frac{1}{2}} I_{n_1}^*$ $(n^{\frac{1}{2}} I_{n_2}^*)$ and $n^{\frac{1}{2}} I_{n_1}$ $(n^{\frac{1}{2}} I_{n_2})$ have the same limiting distribution. The last term in (9.5.9) can be written as

$$\tilde{I}_{n_3}^* = \frac{1}{G_n^*(x,z)} \left(G_n(\beta^*, x, z, z_0) - G_n(\hat\beta, x, z, z_0) \right.$$

$$\left. -G_n(x,z,z_0)\left(\frac{G_n^*(\beta^*,x,z_0)}{G_n^*(x,z_0)} - \frac{G_n(\hat\beta,x,z_0)}{G_n(x,z_0)} \right) \right)$$

$$+ \frac{1}{G_n^*(x,z)} \left(G_n(\hat\beta,x,z,z_0) - G_n(x,z,z_0)\frac{G_n(\hat\beta,x,z_0)}{G_n(x,z_0)} \right)$$

$$=: \frac{1}{G_n^*(x,z)}(I_{n_{31}}^* + G_n(x,z,z_0)I_{n_{32}}^*) + R_n^*$$

$$=: I_{n_3}^* + R_n^*.$$

Parallel to the derivation of (9.5.4) and (9.5.5), it can be shown that

$$\frac{n^{\frac{1}{2}} I_{n_{31}}^*}{G_n^*(x,z)} = \frac{n^{\frac{1}{2}}(\beta^* - \hat\beta)'\dot{G}(\beta,x,z,z_0)}{G(x,z)} + o_p(1),$$

and

$$\frac{n^{\frac{1}{2}} G_n(x,z,z_0)I_{n_{32}}^*}{G_n^*(x,z)} = \frac{n^{\frac{1}{2}}(G_n^*(\beta,x,z_0) - G_n(\beta,x,z_0))G(x,z,z_0)}{G(x,z)G(x,z_0)}$$

$$- \frac{n^{\frac{1}{2}}(G_n^*(x,z_0) - G_n(x,z_0))G(x,z,z_0)G(\beta,x,z_0)}{G(x,z)G(x,z_0)^2}$$

$$- \frac{n^{\frac{1}{2}}(\beta^* - \hat\beta)'\dot{G}(\beta,x,z_0)G(x,z,z_0)}{G(x,z)G(x,z_0)} + o_p(1).$$

By the asymptotic normality of $n^{\frac{1}{2}}(\beta^* - \hat\beta)$ and invoking once again Theorem 2.4 of Giné and Zinn (1990), one can derive that $n^{\frac{1}{2}} I_{n_3}^*$ and $n^{\frac{1}{2}} I_{n_3}$ have the same asymptotic properties. Furthermore,

$$n^{\frac{1}{2}} R_n^* = n^{\frac{1}{2}}\frac{G_n(x,z)}{G_n^*(x,z)}(\hat{A}_n(x,z) - \hat{A}_n(x,z_0)) = n^{\frac{1}{2}}(\hat{A}_n(x,z) - \hat{A}_n(x,z_0)) + o_p(1).$$

From (9.3.1), (9.3.2), (9.5.6), and (9.5.9), we have

$$V_n^*(x,z,z_0) = M_n^*(x,z) - M_n^*(x,z_0)$$

$$= n^{\frac{1}{2}}((A_n^*(x,z) - A_n^*(x,z_0)) - (\hat{A}_n(x,z)) - \hat{A}_n(x,z_0)))$$

$$= n^{\frac{1}{2}}(I_{n_1}^* - I_{n_2}^* + I_{n_3}^*) + o_p(1).$$

Hence $V_n^*(x, z, z_0)$ and $V_n(x, z, z_0)$ have the same limit under H_0. This implies that W_n^* of (9.3.3) converges in distribution to W of (9.2.11). □

Proofs of Theorems 9.3.2 and 9.3.3. Here, we only give the proof of Theorem 9.3.2. Note that both (9.2.2) and (9.2.12) yield the same critical value of the test asymptotically. Hence, Theorem 9.3.3 can be proved in the same fashion.

As to Theorem 9.3.2, we need to prove that (i) the covariance function of V_n^R converges to that of V; (ii) the finite-dimensional distributions of V_n^R converge to those of V; and (iii) the uniform tightness of V_n^R. Zhu, Yuen, and Tang (2002) used similar method for proving the asymptotic validity of a test for a semiparametric random censorship model using the NMCT approach.

Given a sample, the covariance of $V_n^R(x, z, z_0)$ and $V_n^R(x^0, z^0, z_0^0)$ is

$$n^{-1} \sum_{j=1}^{n} f(x, z, z_0, X_j, Z_j; \hat{\beta}, G_n) f(x^0, z^0, z_0^0, X_j, Z_j; \hat{\beta}, G_n),$$

which converges to the covariance function of (9.2.10). Therefore (i) holds. The multivariate central limit theorem implies (ii). Let

$$\mathcal{F} = \{f(x, z, z_0, X, Z; \hat{\beta}, G_n) : (x, z, z_0) \in [0, T] \times [0, \infty)^2\}.$$

For notational convenience, we simply write $f(x_i, z_i, z_{0i}, X, Z; \hat{\beta}, G_n)$ as f_i. Let $P_n(f_i)$ be $n^{-1} \sum_{j=1}^{n} f(x_i, z_i, z_{0i}, X_j, Z_j; \hat{\beta}, G_n)$. Define $[\delta] = \{(f_1, f_2) : f_1, f_2 \in \mathcal{F}, (P_n(f_1 - f_2)^2)^{\frac{1}{2}} \leq \delta\}$. The uniform tightness of (iii) requires that, for each γ and $\epsilon > 0$, there is a $\delta > 0$ such that

$$\limsup_{n \to \infty} P(\sup_{[\delta]} n^{\frac{1}{2}} |V_n^R(x_1, z_1, z_{01}) - V_n^R(x_2, z_2, z_{02})| > \epsilon \mid (X, Z)) \leq \gamma.$$

$$(9.5.10)$$

By Hoeffding inequality,

$$P(n^{\frac{1}{2}} |V_n^R(x_1, z_1, z_{01}) - V_n^R(x_2, z_2, z_{02})| > \epsilon \mid (X, Z))$$
$$\leq 2 \exp \left(\frac{-\epsilon^2}{2 P_n (f_1 - f_2)^2} \right).$$

To derive (9.5.10), we need to show that the covering integral

$$J_2(\delta, \mathcal{F}, P_n) = \int_0^\delta \left(\frac{2 \log N_2(u, \mathcal{F}, P_n)^2}{u} \right)^{\frac{1}{2}} du$$

is finite for some small δ where $N_2(u, \mathcal{F}, P_n)$ is the smallest m for which there exist m functions $f_1, \ldots, f_m$ such that

$$\min_{1 \leq i \leq m} (P_n(f_i - f)^2)^{\frac{1}{2}} \leq u,$$

for any $f \in \mathcal{F}$. Note that $\mathcal{F}$ is a VC subgraph class of functions because f is a linear combination of indicator functions, finite absolutely continuous functions, and their products being members of a VC class. Therefore, for some $c > 0$ and $w > 0$ independent of n,

$$N_2(u, \mathcal{F}, P_n) \leq c\,u^{-w},$$

and then for some $c > 0$

$$J_2(\delta, \mathcal{F}, P_n) \leq c\delta^{\frac{1}{2}}.$$

This implies Condition (16) of the Equicontinuity Lemma in Pollard (1984, pages 150-151). From the Chaining Lemma of Pollard (1984, page 144), there exists a countable dense subset $[\delta]^*$ of $[\delta]$ such that

$$P\left(\sup_{[\delta]^*} n^{\frac{1}{2}} |V_n^R(x_1, z_1, z_{01}) - V_n^R(x_2, z_2, z_{02})| > 26 J_2(\delta, \mathcal{F}, P_n)|(X, Z) \right) \leq 2\delta.$$

Owing to the left continuity of the function f, $[\delta]^*$ can be replaced by $[\delta]$ itself. Choosing δ smaller than $\gamma/2$ and $(\epsilon/(26c))^2$, the tightness of W_n^R is a direct consequence. □

10

Homegeneity Testing for Covariance Matrices

10.1 Introduction

Under a multinormality assumption, hypotheses testing for homogeneity in the k-sample problem can be handled by the likelihood ratio test (LRT). The exact distribution of the LRT is very complicated. When the sample size is sufficiently large, one usually employs the chi-square distribution, the limiting null distribution, for the LRT. Box (1949) obtained a correction factor for Bartlett's LRT and proposed his M statistic with the same chi-square distribution for testing homogeneity in the k-sample problem.

Without the multinormality assumption, likelihood ratios would be different. If one still uses the statistics obtained under normality, then, for example, the asymptotic null distribution for Bartlett's homogeneity test is no longer chi-square, but a linear combination of chi-squares as pointed out by Zhang and Boos (1992). The lack of correct null distribution for traditional statistics forces researchers to look for other means of implementing the tests. Monte Carlo techniques such as the bootstrap represent one resolution. Beran and Srivastava (1985) considered bootstrap implementation of tests based on functions of eigenvalues of a covariance matrix in a one-sample problem. Zhang, Pantula and Boos (1991) proposed a pooled bootstrap methodology. For the k-sample problem, Zhang and Boos (1992, 1993) studied bootstrap procedures to obtain the asymptotic critical values for Bartlett's statistic for homogeneity without the multinormality assumption. Among other things, Zhang and Boos (1993) developed bootstrap theory for quadratic type statistics and demonstrated the idea using Bartlett's test as an example.

In this chapter, we introduce an alternative approach to constructing multivariate tests. It is based on Roy's (1953) union-intersection principle. Most of materials come from Zhu, Ng and Jing (2002). One uses the fact that a random vector is multivariate normal if and only if every non-zero linear function of its elements is univariate normal. This leads to viewing the multivariate hypothesis as the joint statement (intersection) of univariate hypotheses of all linear functions of univariate components, and a joint rejection region con-

sisting of the union of all corresponding univariate rejection regions if they
are available. The two-sample Roy test is in terms of the largest and smallest
eigenvalues of one Wishart matrix in the metric of the other. But, so far, there
is no Roy test for the problem of more than two samples. One reason may
be the difficulty of extending the idea of comparison of variances in terms of
ratio to more than two samples. We briefly describe the difficulty. In a two-
sample case, we may use either σ_1^2/σ_2^2 or σ_2^2/σ_1^2, as they are the reciprocal.
It is not so simple otherwise. If we want an aggregate statistic of pairwise
ratios, one way is to sum up σ_i^2/σ_j^2, $1 \leq i \neq j \leq k$. In case we sum up the
ratios over all $i < j$, as the ratios are not permutation invariant with i and
j, we may obtain conflicting conclusion if we use the sum of the ratios over
$j > i$ as a test statistic. Furthermore if we sum up all ratios over $i \neq j$,
although it will be invariant with i and j, there is some confounding. This
can be demonstrated for $k = 2$ with the statistic $\sigma_1^2/\sigma_2^2 + \sigma_2^2/\sigma_1^2$. When the
first ratio is large, the second will be small, the average will be moderate and
vice versa. It is similar in the general case. However, the absolute values of
the differences $(\sigma_i^2 - \sigma_j^2)/(\sigma_1^2 + \cdots + \sigma_k^2)$, $1 \leq i \neq j \leq k$, are invariant with
respect to i and j. The sum of the absolute values over $i < j$ can be used as a
test statistic without a confounding effect, and so can the maximum of those
absolute values. In this article, we consider both the maximum and the sum
(average), and find in simulations that the sum test statistic works better.

Without multinormality and without reference to the likelihood ratio or
union intersection principle, we obtain homogeneity tests for more than two
samples based on eigenvalues of differences of the sample covariance matrices
subject to a common re-scaling. The asymptotic distributions of the test statistics are identified. We also consider the validity of some resampling techniques,
namely the bootstrap, NMCT and permutation procedures, for calculating the
critical values and p-values for these tests. All of the techniques are asymptotically valid for the problem. There is theory supporting the conclusion that
permutation procedures and NMCT perform better than the bootstrap in
adhering to the nominal level of significance in some cases. Our Monte Carlo
studies indicate that the permutation test generally has higher power than the
bootstrap test and that NMCT is compatible to the bootstrap in power performance. NMCT, if applicable, is easy to implement. Simulation results also
suggest that the test proposed here is better than the bootstrapped Bartlett
test studied by Zhang and Boos (1992).

10.2 Construction of Tests

Let $\boldsymbol{X}_1^{(i)}, \boldsymbol{X}_2^{(i)}, \cdots, \boldsymbol{X}_{m_i}^{(i)}$, $i = 1, \cdots, k$, be an $i.i.d.$ sample from a d-dimensional
distribution with finite fourth moments, mean $\boldsymbol{\mu}^{(i)}$ and covariance matrix
$\boldsymbol{\Sigma}^{(i)}$. We are interested in the homogeneity hypothesis

$$H_0 : \boldsymbol{\Sigma}^{(1)} = \boldsymbol{\Sigma}^{(2)} = \cdots = \boldsymbol{\Sigma}^{(k)} \text{ vs } H_1 : \boldsymbol{\Sigma}^{(i)} \neq \boldsymbol{\Sigma}^{(j)} \text{ for some } i \neq j \,.$$

$$(10.2.1)$$

Denote the sample covariance matrix for the ith sample by

$$\hat{\Sigma}^{(i)} = \frac{1}{m_i} \sum_{j=1}^{m_i} (X_j^{(i)} - \hat{\mu}^{(i)})(X_j^{(i)} - \hat{\mu}^{(i)})^T , \qquad (10.2.2)$$

where $\hat{\mu}^{(i)}$ is either $\mu^{(i)}$ or the sample mean, depending on whether $\mu^{(i)}$ is known or not. The pooled sample covariance matrix is

$$\hat{\Sigma} = \frac{1}{N} \sum_{i=1}^{k} m_i \hat{\Sigma}^{(i)} , \quad \text{where} \quad N = \sum_{i=1}^{k} m_i . \qquad (10.2.3)$$

Based on the idea of multiple comparison, (e.g. see Dunnett (1994), O'Brien (1979, 1981)), we construct tests by combining pairwise comparisons. The pairwise comparison between the lth and ith samples is based on

$$M_{li} = \max \left\{ \text{absolute eigenvalues of } \sqrt{\frac{m_l m_i}{N}} \hat{\Sigma}^{-1/2} (\hat{\Sigma}^{(l)} - \hat{\Sigma}^{(i)}) \hat{\Sigma}^{-1/2} \right\},$$

$$A_{li} = \text{average} \left\{ \text{absolute eigenvalues of } \sqrt{\frac{m_l m_i}{N}} \hat{\Sigma}^{-1/2} (\hat{\Sigma}^{(l)} - \hat{\Sigma}^{(i)}) \hat{\Sigma}^{-1/2} \right\}.$$

$$(10.2.4)$$

We propose using the average of the $k(k-1)/2$ pairwise comparisons as the test statistic,

$$LM = \frac{2}{k(k-1)} \sum_{i<l} M_{li}, \qquad (10.2.5)$$

$$LA = \frac{2}{k(k-1)} \sum_{i<l} A_{li} . \qquad (10.2.6)$$

The null hypothesis is rejected if LM (LA) is greater than the critical value which is to be determined. We first identify the limiting distributions of LM and LA in the following lemma.

To state results, we need some notation for vectorization of a symmetric matrix. For a symmetric $d \times d$ matrix S, let vech(S) be the column vector obtained by stacking up the $d(d+1)/2$ distinct elements of S in the order of the first column vector, then the second column vector omitting the first element, etc.

LEMMA 10.2.1 *Assume $m_i/N \to \lambda_i$, $0 < \lambda_i < 1$, as $m_i \to \infty$ for $i = 1, \cdots, k$, and that the distributions of samples are continuous and have finite fourth moments. Under (10.2.1), the asymptotic joint distribution of $\sqrt{m_l m_i/N} \hat{\Sigma}^{-1/2} (\hat{\Sigma}^{(l)} - \hat{\Sigma}^{(i)}) \hat{\Sigma}^{-1/2}$, $1 \le i, l \le k$, is identical with the asymptotic joint distribution of $\sqrt{\lambda_i} W_l - \sqrt{\lambda_l} W_i$, $1 \le i, l \le k$, where $W_1, \cdots, W_k$*

are independent and vech($\boldsymbol{W}_i$) is multivariate normal with zero mean vector and covariance matrix

$$\boldsymbol{V}_i = cov\big(vech((\boldsymbol{X}_1^{(i)} - \boldsymbol{\mu}^{(i)})(\boldsymbol{X}_1^{(i)} - \boldsymbol{\mu}^{(i)})^T)\big) . \tag{10.2.7}$$

Furthermore, the asymptotic distributions of LM and LA are, respectively, the distributions of the random variables

$$\frac{2}{k(k-1)} \sum_{i<l} \max \big\{ \text{absolute eigenvalues of } \sqrt{\lambda_i}\boldsymbol{W}_l - \sqrt{\lambda_l}\boldsymbol{W}_i \big\}, \tag{10.2.8}$$

$$\frac{2}{k(k-1)} \sum_{i<l} \text{average} \big\{ \text{absolute eigenvalues of } \sqrt{\lambda_i}\boldsymbol{W}_l - \sqrt{\lambda_l}\boldsymbol{W}_i \big\} .$$

$$\tag{10.2.9}$$

Under the alternative in (10.2.1), LM and LA diverge to infinity.

Although the conclusion above does not lend itself to the calculation of p-values, we may employ resampling techniques for implementation.

10.3 Monte Carlo Approximations

We consider three sampling techniques in this section, including the bootstrap, NMCT and the permutation test.

10.3.1 Classical Bootstrap

We follow the pooled re-sampling procedure suggested by Zhang and Boos (1992) and let

$$(\boldsymbol{Z}_1, \cdots, \boldsymbol{Z}_N) = \big(\boldsymbol{X}_1^{(1)} - \hat{\boldsymbol{\mu}}^{(1)}, \cdots, \boldsymbol{X}_{m_1}^{(1)} - \hat{\boldsymbol{\mu}}^{(1)}, \cdots, \boldsymbol{X}_1^{(k)} - \hat{\boldsymbol{\mu}}^{(k)}, \cdots, \boldsymbol{X}_{m_k}^{(k)} - \hat{\boldsymbol{\mu}}^{(k)}\big) , \tag{10.3.1}$$

where $\hat{\boldsymbol{\mu}}^{(i)}$ is either $\boldsymbol{\mu}^{(i)}$ or the sample mean, depending on whether $\boldsymbol{\mu}^{(i)}$ is known or not. Let $(\boldsymbol{Z}_1^*, \cdots, \boldsymbol{Z}_N^*)$ be drawn with replacement from the given sample $(\boldsymbol{Z}_1, \cdots, \boldsymbol{Z}_N)$, and let

$$\hat{\boldsymbol{\Sigma}}_i^* = \frac{1}{m_i} \sum_{j=1+N_{i-1}}^{N_i} (\boldsymbol{Z}_j^* - \bar{\boldsymbol{Z}}^{i*})(\boldsymbol{Z}_j^* - \bar{\boldsymbol{Z}}^{i*})^T , \quad i = 1, \cdots, k , \tag{10.3.2}$$

where $\bar{\boldsymbol{Z}}^{i*}$ is the sample mean of $\boldsymbol{Z}^*{}_j$ for $N_{i-1} + 1 \leq N_i$, $N_i = \sum_{l=1}^i m_l$ for $i = 1, \cdots, k$, and $N_0 = 0$. Let

$$M_{li}^B = \max \bigg\{ \text{absolute eigenvalues of } \sqrt{\frac{m_l m_i}{N}} \hat{\boldsymbol{\Sigma}}^{-1/2} \big(\hat{\boldsymbol{\Sigma}}_l^* - \hat{\boldsymbol{\Sigma}}_i^*\big) \hat{\boldsymbol{\Sigma}}^{-1/2} \bigg\},$$

$$A_{li}^B = \text{average} \bigg\{ \text{absolute eigenvalues of } \sqrt{\frac{m_l m_i}{N}} \hat{\boldsymbol{\Sigma}}^{-1/2} \big(\hat{\boldsymbol{\Sigma}}_l^* - \hat{\boldsymbol{\Sigma}}_i^*\big) \hat{\boldsymbol{\Sigma}}^{-1/2} \bigg\}.$$

The bootstrap counterparts of (10.2.5) and (10.2.6) are then

$$LM_B = \frac{2}{k(k-1)} \sum_{i<l} M_{li}^B, \qquad (10.3.3)$$

$$LA_B = \frac{2}{k(k-1)} \sum_{i<l} A_{li}^B. \qquad (10.3.4)$$

The asymptotic equivalence of LM_B and LM and of LA_B and LA is established in the following theorem.

THEOREM 10.3.1 *Assume the conditions in Lemma 10.2.1. For almost all sequences* $\left(\boldsymbol{X}_1^{(1)}, \cdots, \boldsymbol{X}_{m_1}^{(1)}, \cdots; \boldsymbol{X}_1^{(2)}, \cdots, \boldsymbol{X}_{m_2}^{(2)}, \cdots; \cdots; \boldsymbol{X}_1^{(k)}, \cdots, \boldsymbol{X}_{m_k}^{(k)}, \cdots\right)$, *of independent* $d \times 1$ *random vectors having finite fourth moments with* $E(\boldsymbol{X}_j^{(i)}) = \boldsymbol{\mu}^{(i)}$ *and* $cov(\boldsymbol{X}_j^{(i)}) = \boldsymbol{\Sigma}$ *for* $i = 1, \cdots, k$, *the conditional distribution of* LM_B (LA_B) *given the samples* $\left(\boldsymbol{X}_1^{(1)}, \cdots, \boldsymbol{X}_{m_1}^{(1)}; \boldsymbol{X}_1^{(2)}, \cdots, \boldsymbol{X}_{m_2}^{(2)}, \cdots; \right.$ $\left.\boldsymbol{X}_1^{(k)}, \cdots, \boldsymbol{X}_{m_k}^{(k)}\right)$ *converges to the unconditional asymptotic distribution of* LM *(LA).*

In view of this asymptotic equivalence, the critical value of LM (LA) for testing H_0 can be calculated by repeated bootstrap sampling from the given sample data.

10.3.2 NMCT Approximation

When the k samples have the same size, say m, we suggest another conditional test procedure which is much easier to implement. The motivation of the method is given below. We also give a brief justification for the exact validity of the NMCT test in a special case. The asymptotic validity will be stated as a theorem.

Consider the two-sample case as an illustration. Suppose that under the null hypothesis all variables $(\boldsymbol{X}_1^{(1)}, \boldsymbol{X}_2^{(1)}, \cdots, \boldsymbol{X}_m^{(1)})$ and $(\boldsymbol{X}_1^{(2)}, \boldsymbol{X}_2^{(2)}, \cdots, \boldsymbol{X}_m^{(2)})$ are *i.i.d.* from a d-dimensional distribution with a given mean. Without loss of generality, assume the mean to be zero. Let Y_j denote $(\boldsymbol{X}_j^{(1)})(\boldsymbol{X}_j^{(1)})^T - (\boldsymbol{X}_j^{(2)})(\boldsymbol{X}_j^{(2)})^T$. By assumption, Y_j has a symmetric distribution. For a random sign e_j independent of Y_j, Y_j and $e_j Y_j$ are identical in distribution and e_j is independent of $e_j Y_j$. The latter assertion can be seen by invoking the independence of e_j and Y_j and the symmetry of Y_j. Therefore, for any statistic $T(Y_1, \cdots, Y_m)$, its distribution is the same as that of $T(e_1 Y_1, \cdots, e_m Y_m)$ where e_i's are *i.i.d.* random signs. Consequently, generate r sets of Rademacher variables $(e_1, \cdots, e_m)$, and then obtain r values of $T(e_1 Y_1, \cdots, e_m Y_m)$, say $T^1, \cdots T^r$. Denote the value of the original T as T^0. We know that T^i, $i = 0, 1, \cdots, r$, are $r + 1$ *i.i.d.* variables. Suppose for the moment the null hypothesis will be rejected for large value of T (for two-sided tests, modifications are easily done). The p-value can be estimated by the fraction of values in $T^0, T^1, \cdots, T^r$ that are larger than or equal to T^0. If the estimated p-value is smaller than the nominal level α, the null hypothesis will be rejected. This explains the exact validity of the NMCT approximation.

In practice, one cannot assume that variables in different samples are *iid* and the mean is known. We have to use the estimate in place of unknown mean. In the following we give the detail of constructing tests and of the consistency of the NMCT approximation for the general case.

Since the NMCT is also a conditional test, we can work with the standardized data as in the bootstrap procedure:

$$Z_j^{(i)} = \hat{\Sigma}^{-1/2}\left(X_j^{(i)} - \hat{\mu}^{(i)}\right), \quad j = 1, \cdots, m; \; i = 1, \cdots, k . \qquad (10.3.5)$$

Let $\{e_1, \cdots, e_m\}$ be a set of random signs, the NMCT of $\hat{\Sigma}^{-1/2}(\hat{\Sigma}^{(l)} - \hat{\Sigma}^{(i)})\hat{\Sigma}^{-1/2}$ is, $1 \leq i < l \leq k$,

$$W_{li} = \frac{1}{m}\sum_{j=1}^{m} e_j\left[Z_j^{(l)}(Z_j^{(l)})^T - Z_j^{(i)}(Z_j^{(i)})^T\right] . \qquad (10.3.6)$$

The NMCT counterparts of LM and LA are

$$LM_R = \frac{2}{k(k-1)}\sum_{i<l}\max\left\{\text{absolute eigenvalues of } \sqrt{\frac{m^2}{N}}W_{li}\right\}, \quad (10.3.7)$$

$$LA_R = \frac{2}{k(k-1)}\sum_{i<l}\text{average}\left\{\text{absolute eigenvalues of } \sqrt{\frac{m^2}{N}}W_{li}\right\} . \qquad (10.3.8)$$

We need to verify that LM_R (LA_R) is asymptotically equivalent to LM (LA).

THEOREM 10.3.2 *Under the assumptions of Lemma 10.2.1, the conditional distribution of LM_R (LA_R) given the data converges to the unconditional asymptotic distribution of LM (LA).*

The p-value is estimated similar as at the end of Section 10.3.1. Let $LM_R^{(1)}, \cdots, LM_R^{(r)}$ be r replications of NMCT with r independent sets of random signs and let $LM_R^{(0)}$ be the value of the original test statistic LM. The estimated p-value equals the fraction of the values which are greater than or equal to $LM_R^{(0)}$. The same procedure can be applied to LA_R.

10.3.3 Permutation Test

A drawback of the NMCT is its restriction to equal sample size. The permutation test can be applied to samples of unequal sizes. It also has some advantages over the bootstrap, but is harder to implement than NMCT. It is easy to see that, similar to NMCT, when all variables in samples are *iid*, the exact validity of the permutation tests can be achieved. The justification is similar to that described for NMCT.

Pool the standardized data

$$\hat{\boldsymbol{\Sigma}}^{-1/2}\big(\boldsymbol{X}_j^{(i)} - \hat{\boldsymbol{\mu}}^{(i)}\big), \quad j = 1, \cdots, m_i; \; i = 1, \cdots, k, \qquad (10.3.9)$$

into a sample of size N, then randomly divide it into k samples such that the ith sample has size m_i. Denote the ith sample by $\boldsymbol{Z}_j^{(i)}$, $j = 1, \cdots, m_i$, and let

$$\hat{\boldsymbol{\Sigma}}_P^{(i)} = \frac{1}{m_i} \sum_{j=1}^{m_i} \boldsymbol{Z}_j^{(i)}(\boldsymbol{Z}_j^{(i)})^T. \qquad (10.3.10)$$

The permutation test statistics are

$$LM_P = \frac{2\sum_{i<l} \max\left\{\text{absolute eigenvalues of } \sqrt{\frac{m_l m_i}{N}}\big(\hat{\boldsymbol{\Sigma}}_P^{(l)} - \hat{\boldsymbol{\Sigma}}_P^{(i)}\big)\right\}}{k(k-1)},$$

$$(10.3.11)$$

$$LM_P = \frac{2\sum_{i<l} \text{average}\left\{\text{absolute eigenvalues of } \sqrt{\frac{m_l m_i}{N}}\big(\hat{\boldsymbol{\Sigma}}_P^{(l)} - \hat{\boldsymbol{\Sigma}}_P^{(i)}\big)\right\}}{k(k-1)}.$$

$$(10.3.12)$$

Analogous to NMCT, the exact validity of the permutation tests for the case of given means can be obtained. In fact, under the null hypothesis, the permutation counterpart has the same distribution as that of the original test statistic. Therefore, similar to that illustrated for NMCT, exact validity can be expected. As with NMCT it is of course restrictive, but simulation studies show that in unknown mean cases, permutation tests outperform bootstrap tests in getting closer to the nominal level. The following covers the asymptotic validity of permutation tests for the general case.

THEOREM 10.3.3 *Under the assumptions of Lemma 10.2.1, the conditional distribution of LM_P (LA_P) given the data converges to the unconditional asymptotic distribution of LM (LA).*

With r independent random permutations, we have r replications of (10.3.11), $LM_P^{(1)}, \cdots, LM_P^{(r)}$. The p-value can be estimated as in preceding procedures.

10.3.4 Simulation Study

This section reports the results of some Monte Carlo studies. These are carried out to compare the three procedures using three families of multivariate distributions: multinormal $N(\mathbf{0}, \mathbf{I}_d)$, multivariate t-distribution $MT(5; \mathbf{0}, \mathbf{I}_d)$, and a contaminated normal distribution $NC_2(\mathbf{0}, \mathbf{I}_d)$ whose components are independent, each being $N(0, 1)$ with probability 0.9 and a $\chi^2_{(2)}$ with probability 0.1. We consider $k = 2$ and $k = 6$, and the dimension of random vector $d = 2$ and $d = 5$. The nominal 5% level of significance is chosen. In each procedure, the number of replications for calculating a critical value is $r = 500$. Each actual proportion of rejections of H_0 is based on 1000 simulations. As expected, the tests perform better when the means are known. Here we only report results relating to the case of an unknown mean. As one can see from Table 10.1, most of the time the actual proportion of rejection by permutation (PERM) is closer to the nominal α than is the classical bootstrap (BOOT), and in this aspect NMCT is comparable to the bootstrap (BOOT). For different sample sizes, where NMCT is not available, the results are given in Table 10.2. The table shows that PERM is better than BOOT in 8 of 12 simulations. Comparing LA with LM, we found that with equal sample sizes, when $k = 2$, LA is worse than LM most of the time; when $k = 6$, LA is better in all cases. With different sample sizes, LA is better than LM most of the time.

Table 10.1: Percentage of times H_0 (10.2.1) was rejected

$k = 2, m_1 = m_2 = 20$

		$N(\mathbf{0}, \mathbf{I}_d)$			$MT(5; \mathbf{0}, \mathbf{I}_d)$			$NC_2(\mathbf{0}, \mathbf{I}_d)$		
		NMCT	BOOT	PERM	NMCT	BOOT	PERM	NMCT	BOOT	PERM
$d = 2$	LM	0.053	0.045	0.048	0.053	0.046	0.054	0.055	0.046	0.056
	LA	0.059	0.057	0.054	0.062	0.046	0.057	0.063	0.060	0.061
$d = 5$	LM	0.055	0.045	0.056	0.050	0.044	0.051	0.057	0.039	0.053
	LA	0.054	0.047	0.054	0.059	0.031	0.060	0.063	0.041	0.059

$k = 6, m_i = 20, i = 1, \cdots, 6$

		$N(\mathbf{0}, \mathbf{I}_d)$			$MT(5; \mathbf{0}, \mathbf{I}_d)$			$NC_2(\mathbf{0}, \mathbf{I}_d)$		
		NMCT	BOOT	PERM	NMCT	BOOT	PERM	NMCT	BOOT	PERM
$d = 2$	LM	0.033	0.062	0.060	0.032	0.064	0.059	0.033	0.061	0.058
	LA	0.053	0.058	0.057	0.060	0.060	0.056	0.057	0.060	0.055
$d = 5$	LM	0.040	0.063	0.059	0.043	0.058	0.056	0.043	0.057	0.063
	LA	0.056	0.060	0.054	0.049	0.045	0.053	0.055	0.054	0.055

The bootstrap results in these Monte Carlo studies also provide some evidence for comparing the test statistics LM and LA of (10.3.6) with Bartlett's statistic. As shown in Zhang and Boos (1992, p.428), the bootstrap procedure for Bartlett's homogeneity test performs worse as dimension d (they used p for dimension) increases. The bootstrap procedure of our tests is quite stable; see Table 10.1. When $m_1 = m_2 = 20$, $\alpha = 0.05$ and $d = 2$, our proportions of rejections for three distributions are 0.045, 0.046 and 0.046, against their 0.046, 0.045 and 0.50 respectively. But when dimension increases to $d = 5$, ours become 0.045, 0.044 and 0.039, against their 0.012, 0.023 and 0.019 respectively.

Table 10.2: Percentage of times H_0 (10.2.1) was rejected

$k = 2, m_1 = 20, m_2 = 40$

		$N(\mathbf{0}, \boldsymbol{I}_d)$		$MT(5; \mathbf{0}, \boldsymbol{I}_d)$		$NC_2(\mathbf{0}, \boldsymbol{I}_d)$	
		BOOT	PERM	BOOT	PERM	BOOT	PERM
$d = 2$	LM	0.042	0.041	0.046	0.045	0.058	0.055
	LA	0.054	0.053	0.058	0.056	0.054	0.053
$d = 6$	LM	0.047	0.049	0.053	0.055	0.064	0.060
	LA	0.055	0.052	0.049	0.048	0.053	0.053

$k = 6, m_1 = m_2 = 20, m_3 = m_4 = 30, m_5 = m_6 = 40$

		$N(\mathbf{0}, \boldsymbol{I}_d)$		$MT(5; \mathbf{0}, \boldsymbol{I}_d)$		NC	
		BOOT	PERM	BOOT	PERM	BOOT	PERM
$d = 2$	LM	0.060	0.057	0.045	0.043	0.054	0.052
	LA	0.056	0.055	0.041	0.045	0.055	0.054
$d = 6$	LM	0.066	0.062	0.065	0.063	0.064	0.059
	LA	0.060	0.057	0.043	0.045	0.045	0.055

The power of the tests was also studied for $k = 2$ samples with sample size $m_1 = m_2 = 20$, for dimension $d = 2$. Multinormal and multivariate-t distributions, $N(\boldsymbol{\mu}, \boldsymbol{\Sigma})$ and $MT(5, 0, \boldsymbol{\Sigma})$, were used to generate data. We pair the identity matrix $\boldsymbol{I}_2$ with $\boldsymbol{C}_2$ and with $\boldsymbol{V}_2$, respectively, where

$$C_2 = \begin{pmatrix} 1 & 0.5 \\ 0.5 & 1 \end{pmatrix}, \quad V_2 = \begin{pmatrix} 2 & 0 \\ 0 & 4 \end{pmatrix}.$$

The results are given in Table 10.3. For comparison with Bartlett's test as studied by Zhang and Boos (1992), we also calculate the adjusted power. The rows marked $(a - p)$ are obtained by using as critical values the 5th percentile of the empirical distribution of p values under H_0 obtained in constructing Table 10.1. The table shows that the bootstrap has higher power when the distribution is normal, but lower power than the permutation test and NMCT in the case of multivariate t. This is also true for simulation results in the three-dimensional case, not shown here in order to save space. Furthermore, LA has better performance than LM most of the time.

The power studies of the bootstrap are in favor of our tests. Zhang and Boos (1992) performed power studies of the bootstrapped Bartlett's test with C_2 and V_2. The corresponding values (BartlettB) in Table 2 of Zhang and Boos (1992) and in our Table 10.3 are collated below for easier comparison $(\alpha = 0.05, k = 2, d = 2, m_1 = m_2 = 20)$ where the first column is for the case of $N(0, \boldsymbol{V}_2)$ against $N(0, \boldsymbol{I}_2)$, the second column for $MT(5, 0, \boldsymbol{V}_2)$ against $MT(5, 0, \boldsymbol{I}_2)$, the third column for $N(0, \boldsymbol{C}_2)$ against $N(0, \boldsymbol{I}_2)$, and the fourth column for $MT(5, 0, \boldsymbol{C}_2)$ against $MT(5, 0, \boldsymbol{I}_2)$. The values in parentheses are the adjusted powers, assuming the population means to be unknown parameters:

BartlettB   0.642(0.657) 0.487(0.525) 0.233(0.243) 0.155(0.177)
BOOT of (10.2.6) 0.770(0.818) 0.538(0.617) 0.276(0.299) 0.229(0.270)
NMCT of (10.2.6) 0.784(0.817) 0.570(0.640) 0.256(0.296) 0.230(0.274)
PERM of (10.2.6) 0.770(0.814) 0.589(0.637) 0.265(0.297) 0.234(0.275)

Table 10.3: Power study for $d = 2$, $k = 2$, $m_1 = m_2 = 20$

	$N(\mathbf{0}, \mathbf{C_2})$ & $N(\mathbf{0}, \mathbf{I_2})$			$N(\mathbf{0}, \mathbf{V_2})$ & $N(\mathbf{0}, \mathbf{I_2})$		
	NMCT	BOOT	PERM	NMCT	BOOT	PERM
LM	0.193	0.227	0.193	0.753	0.781	0.762
$LM(a-p)$	0.229	0.246	0.223	0.815	0.825	0.817
LA	0.256	0.276	0.265	0.784	0.770	0.770
$LA(a-p)$	0.296	0.299	0.297	0.817	0.818	0.814
	$MT(5; \mathbf{0}, \mathbf{C_2})$ & $MT(5; \mathbf{0}, \mathbf{I_2})$			$MT(5; \mathbf{0}, \mathbf{V_2})$ & $Mt(5; \mathbf{0}, \mathbf{I_2})$		
	NMCT	BOOT	PERM	NMCT	BOOT	PERM
LM	0.228	0.231	0.233	0.553	0.521	0.561
$LM(a-p)$	0.261	0.262	0.265	0.636	0.604	0.823
LA	0.230	0.229	0.234	0.570	0.538	0.589
$LA(a-p)$	0.274	0.270	0.275	0.640	0.617	0.637

In summary, we have the following recommendations: (1) the average value test outperforms the maximum value test; (2) the tests of (10.2.5) and (10.2.6) are preferred over Bartlett's test, with or without the bootstrap; (3) if sample sizes are equal, use the NMCT procedure for easier implementation even though its power performance may be slightly worse; (4) if sample sizes are not equal, the permutation procedure is a good choice.

10.4 Appendix

To simplify notation we rewrite, in the two-sample case, m_1 as m and m_2 as n, and the second sample as $(\mathbf{Y}_1, \cdots, \mathbf{Y}_n)$. In this way, $N = m + n$. It is clear that the convergence of the test statistics follows the convergence of the random matrices defined. Hence we deal with convergence of random matrices. Furthermore note that in the k-sample case the estimate $\hat{\mathbf{\Sigma}}$ of the covariance matrix based on all data converges to a constant matrix in probability and will not affect the limiting behavior of the test statistics. Hence, we simply regard it as an identity matrix when studying the limit properties of tests.

The proof of Lemma 10.2.1 is simple and the proof of Theorem 10.3.1 is a direct application of Theorem 2.4 in Giné and Zinn (1990). The details are omiited.

Proof of Theorem 10.3.2. We have to first prove the asymptotic normality of $\{\sqrt{m_i m_l/N}\boldsymbol{W}_{li}, 1 \leq i < l \leq k\}$. We need only prove the asymptotic normality of all linear combinations of the matrices $\sqrt{m_i m_l/N}\boldsymbol{W}_{li}$ having asymptotically the same covariance structure as that of the limiting random matrices in Theorem 10.3.1. That is, for any constants b_{il} with at least one being nonzero, $\sum_{1\leq i<l\leq k} b_{il}\sqrt{m_i m_l/N}\boldsymbol{W}_{li}$ is asymptotically normal in the sense of Lemma 10.2.1. These can be derived by the above with some more calculation, details begin omitted. The proof is completed. □

Proof of Theorem 10.3.3. We first prove the convergence of the permutation empirical process in the two-sample case. Write m_1 as m and the second sample as $\{\boldsymbol{Y}_1,\cdots,\boldsymbol{Y}_n\}$, $N = m + n$. Let F_m and F_m^P be the empirical distributions based on $\{\boldsymbol{X}_1,\cdots,\boldsymbol{X}_m\}$ and $\{\boldsymbol{Z}_1,\cdots,\boldsymbol{Z}_m\}$ respectively, and G_n and G_n^P the empirical distributions based on $\{\boldsymbol{Y}_1,\cdots,\boldsymbol{Y}_n\}$ and $\{\boldsymbol{Z}_{m+1},\cdots,\boldsymbol{Z}_{m+n}\}$. Further, let $H_N(t) = (m/N)F_m(t) + (n/N)G_n(t)$ and $H(t) = \lambda F(t) + (1-\lambda)G(t)$. Applying Theorem 1 of Præstgaard (1995, p. 309), for almost all series $\{\boldsymbol{X}_i\}$ and $\{\boldsymbol{Y}_i\}$,

$$\{\sqrt{nm/N}(F_m^P(t) - G_n^P(t)) : t \in R^1\}$$
$$= \{\sqrt{mN/n}(F_m^P(t) - H_N(t)) : t \in R^1\}$$
$$\Longrightarrow RV_H =: \{RV_H(t) : t \in R^1\}, \tag{10.4.1}$$

where " $\Longrightarrow$ " stands for the convergence in distribution, RV_H is a P-Brownian bridge with $H = \lambda F + (1-\lambda)G$. The convergence is convergence in distribution in $l^\infty(\mathcal{F})$, consisting of bounded, real-valued functions defined on $\mathcal{F}$, the class of indicator functions of half spaces $\{\boldsymbol{a}^\tau \cdot \leq t\}$. As usual, (see, e.g., Giné and Zinn (1984)), the supremum norm on this space is considered. Note that all sample paths of RV_H are contained in $C(\mathcal{F}, H)$, a sub-collection consisting of all bounded, uniformly continuous functions under the $L^2(H)$-seminorm $d^2(f,g) = E_H(f-g)^2 - (E_H(f-g))^2$. It is known that $C(\mathcal{F}, H)$ is separable (e.g. see Pollard (1984), p.169, ex.7). Furthermore, any point in $C(\mathcal{F}, H)$ can easily be showed to be completely regular (Pollard (1984), p.67). By the representation theory (e.g. Pollard (1984), p.71), we have, under uniform norm,

$$\{\sqrt{mN/n}(F_m^P(t) - H_N(t)) : t \in R^1\} \longrightarrow \{RV_H(t) : t \in R^1\} \quad a.s. \tag{10.4.2}$$

We now turn to the proof for $\sqrt{mn/N}\{(\hat{\boldsymbol{\Sigma}}_P^{(1)} - \hat{\boldsymbol{\Sigma}}_P^{(2)})$. Consider the upper-left element on the diagonal of the matrix, $\sqrt{mn/N}\{1/m\sum_{i=1}^m[(Z_i^P)^2 - 1/n\sum_{i=1}^n(Z_{i+m}^P)^2]\}$. Write it as

$$\sqrt{mn/N}\int(t^2 - \sigma^2)d\left(F_m^P(t) - G_n^P(t)\right)$$
$$= \int(t^2 - \sigma^2)d\{\sqrt{mN/n}(F_m^P(t) - H_N(t))\}.$$

From (10.4.2), it converges to $T^P := \int (t^2 - \sigma^2) d\,RV_H(t)$ a.s. This stochastic integral is distributed normally. The work remaining is to check that its variance coincides with the variance of the upper-left element of $\boldsymbol{W}_1$, $E(x^2 - \sigma^2)^2$. Note that under the condition of Theorem 10.3.3, $H = F$. Via some elementary calculations, we have

$$
\begin{aligned}
E((T^P)^2) &= E\Big(\int (t^2 - \sigma^2)(t_1^2 - \sigma^2) d\,RV_H(t) d\,RV_H(t_1)\Big) \\
&= \int (t^2 - \sigma^2)^2 E(d\,RV_H(t))^2) \\
&= \int (t^2 - \sigma^2)^2 d\,F(t) = E(x^2 - \sigma^2)^2. \tag{10.4.3}
\end{aligned}
$$

The third equation uses $\int (t^2 - \sigma^2)^2 (d\,F(t))^2 = 0$.

For the general case, we start with a lemma. We consider only the 3-sample case, more samples can be treated with more complicated calculations. Let the data be $\{\boldsymbol{X}_1^{(1)}, \cdots, \boldsymbol{X}_{m_1}^{(1)}, \boldsymbol{X}_1^{(2)}, \cdots, \boldsymbol{X}_{m_2}^{(2)}, \boldsymbol{X}_1^{(3)}, \cdots, \boldsymbol{X}_{m_3}^{(3)}\}$ and let $\{\boldsymbol{Z}_1^{(1)}, \cdots, \boldsymbol{Z}_{m_1}^{(1)}, \boldsymbol{Z}_1^{(2)}, \cdots, \boldsymbol{Z}_{m_2}^{(2)}, \boldsymbol{Z}_1^{(3)}, \cdots, \boldsymbol{Z}_{m_3}^{(3)}\}$ be the data generated by permutation. Let $N = m_1 + m_2 + m_3$. Further, for $l = 1, 2, 3$ let F_{m_l} and $F_{m_l}^P$ be, respectively, the empirical distributions based on $\{\boldsymbol{X}_1^{(l)}, \cdots, \boldsymbol{X}_{m_l}^{(l)}\}$, and $\{\boldsymbol{Z}_1^{(l)}, \cdots, \boldsymbol{Z}_{m_l}^{(l)}\}$, and let $H_{N-m_1}(t) = (m_2/(N - m_1))F_{m_2}(t) + (m_3/(N - m_1))F_{m_3}(t)$.

LEMMA 10.4.1 *Under the conditions of Theorem 10.3.3, the conditional empirical process $\{\sqrt{m_2(N - m_1)/m_3}(F_{m_2}^P(t) - H_{N-m_1}(t)) : t \in R^1\}$ given $\{\boldsymbol{Z}_1^{(1)}, \cdots, \boldsymbol{Z}_{m_1}^{(1)}\}$ converges weakly to $\{RV_F(t) : t \in R^1\}$, where F is the distribution of the random variable $\boldsymbol{X}$.*

Proof. Note that when $\{\boldsymbol{Z}_1^{(1)}, \cdots, \boldsymbol{Z}_{m_1}^{(1)}\}$ is given, the process $\{\sqrt{m_2(N - m_1)/m_3}\,(F_{m_2}^P(t) - H_{N-m_1}(t)) : t \in R^1\}$ is almost the same as that in (10.4.2). Following the arguments used in the proof of Theorem 1 of Præstgaard (1995), we can derive the conclusion. Details are omitted.

We now turn to the proof of the theorem and first consider the asymptotic normality of $\{(\hat{\boldsymbol{\Sigma}}_P^{(i)} - \hat{\boldsymbol{\Sigma}}_P^{(l)}), 1 \leq i < l \leq 3\}$. We show, for any constants b_{il} with at least one being nonzero, $\sum_{1 \leq i < l \leq 3} b_{il} \sqrt{m_i m_l / N}(\hat{\boldsymbol{\Sigma}}_P^{(i)} - \hat{\boldsymbol{\Sigma}}_P^{(l)})$ is asymptotically normal in the sense of Lemma 10.2.1. In the 3-sample case, we consider the empirical permutation process $\sum_{1 \leq i < l \leq 3} b_{il} \sqrt{m_i m_l / N}(F_{m_i i}^P - F_{m_l l}^P)$ first, convergence of $\{(\hat{\boldsymbol{\Sigma}}_P^{(i)} - \hat{\boldsymbol{\Sigma}}_P^{(l)}), 1 \leq i < l \leq 3\}$ will be a consequence. Invoking $F_{m_3 3}^P = (N F_N - m_1 F_{m_1}^P - m_2 F_{m_2}^P)/m_3$, it can be verified that

$$\sum_{1 \le i < l \le 3} b_{il} \sqrt{m_i m_l / N} (F^P_{m_i} - F^P_{m_l})$$

$$= \sqrt{m_1} \Big(b_{12} \sqrt{m_2/N} + b_{13}(1 + m_1/m_3) \sqrt{m_3/N}$$

$$+ b_{23} \sqrt{m_1 m_2 / (m_3 N)} \Big) (F^P_{m_1} - F_N)$$

$$+ \sqrt{m_2} \Big(-b_{12} \sqrt{m_1/N} + b_{23}(1 + m_2/m_3) \sqrt{m_3/N}$$

$$+ b_{13} \sqrt{m_1 m_2 / (m_3 N)} \Big) (F^P_{m_2} - F_N)$$

$$=: \sqrt{m_1} b_{n1} (F^P_{m_1} - F_N) + \sqrt{m_2} b_{n2} (F^P_{m_2} - F_N) \qquad (10.4.4)$$

where b_{n1} and b_{n2} go to constants. Further, noting that $F_N = m_1/N(F^P_{m_1} - F^P_{N-m_1}) + F^P_{N-m_1}$,

$$\sqrt{m_1} b_{n1} (F^P_{m_1} - F_N) + \sqrt{m_2} b_{n2} (F^P_{m_2} - F_N)$$

$$= \sqrt{m_1} (b_{n1} + b_{n2} \sqrt{m_1 m_2}/N)(F^P_{m_1} - F_N) + \sqrt{m_2} b_{n2} (F^P_{m_2} - F^P_{N-m_1}). \qquad (10.4.5)$$

Note that $(F^P_{m_1} - F_N)$ is conditionally independent of $(F^P_{m_2} - F^P_{N-m_1})$. Together with the proof for the two-sample case and the lemma, the process then converges weakly to a Gaussian process. The work remaining is to check that the limiting covariance structure of $\sum_{1 \le i < l \le 3} b_{il} \sqrt{m_i m_l / N} (F_{m_i i} - F_{m_l l})$ coincides with that of $\sum_{1 \le i < l \le 3} b_{il} \sqrt{m_i m_l / N} (F^P_{m_i i} - F^P_{m_l l})$. As above,

$$\sum_{1 \le i < l \le 3} b_{il} \sqrt{m_i m_l / N} (F_{m_i} - F_{m_l})$$

$$= \sqrt{m_1} (b_{n1} + b_{N2} \sqrt{m_1 m_2}/N)(F_{m_1} - F_N) + \sqrt{m_2} b_{n2} (F_{m_2} - F_{N-m_1}). \qquad (10.4.6)$$

It is enough to show that

1) $\sqrt{m_1 N/(N - m_1)} (F^P_{m_1} - F_N)$ and $\sqrt{m_1 (N - m_1)/m_3} (F^P_{m_2} - F^P_{N-m_1})$ have the same limiting covariance structures as those of

$$\sqrt{m_1 N/(N - m_1)} (F_{m_1} - F_N) \quad \text{and} \quad \sqrt{m_1 (N - m_1)/m_3} (F_{m_2} - F_{N-m_1})$$

respectively, and

2) $F_{m_1} - F_N$ is uncorrelated with $F_{m_2} - F_{N-m_1}$.

For 1), noting that $F_{m_1} - F_N = (N - m_1)/N(F_{m_1} - F_{N-m_1})$, it is easy to see that the covariance at (t, t_1) is

$$R(t, t_1) = \frac{m_1(N - m_1)}{N} \Big(\frac{1}{m_1} + \frac{1}{N - m_1} \Big) (F(t \wedge t_1) - F(t)F(t_1))$$

$$= F(t \wedge t_1) - F(t)F(t_1), \qquad (10.4.7)$$

the covariance structure of a P-Brownian bridge, where "$\wedge$" denotes minimum; similarly for $\sqrt{m_1(N - m_1)/m_3}(F_{m_2} - F_{N-m_1})$.

For 2), via elementary calculations we have, applying the independence of the variables having the common distribution F,

$$E(F_{m_1}(t) - F_N(t))(F_{m_2}(t_1) - F_{N-m_1}(t_1))$$
$$= -\frac{m_3}{N} E(F_{N-m_1}(t) - F(t))(F_{m_2}(t_1) - F_{m_3}(t_1))$$
$$= -\frac{m_3}{(N(N - m_1))} [(F(t \wedge t_1) - F(t)F(t_1)) - (F(t \wedge t_1) - F(t)F(t_1))] = 0.$$

The proof is complete. $\square$

References

Aert, M., Claeskens, G., and Hart, J. D. (1999). Testing lack of fit in multiple regression. *J. Amer. Statist. Assoc.* **94**, 869-879.

Aki, S. (1987). On non-parametric tests for symmetry. *Ann. Inst. Statist. Math.* **39**, 457-472.

Aki, S. (1993). On non-parametric tests for symmetry in R^m. *Ann. Inst. Statist. Math.* **45**, 787-800.

Aly, E. A. A., Kochar, S. C. and McKeague, I. W. (1994). Some tests for comparing cumulative incidence functions and cause-specific hazard rates. *J. Am. Statist. Assoc.* **89**, 994-999.

Anderson, T. W. (1984). *An Introduction to Multivariate Statistical Analysis.* John Wiley, New York

Antille, L., Kersting, G., and Zucchini, W. (1982). Testing symmetry. *J. Amer. Statist. Assoc.* **77**, 639-646.

Aras, G. and Deshpandé, J. V. (1992). Statistical analysis of dependent competing risks. *Statist. Decis.* **10**, 323-336.

Azzalini, A., Bowman, A. W. and Härdle, W. (1989). On the use of nonparametric regression for model checking. *Biometrika* **76** 1-11.

Bagai, I., Deshpande, J. V. and Kochar, S. C. (1989). Distribution free tests for stochastic ordering in the competing risks model. *Biometrika* **76**, 775–781.

Bagai, I., Deshpande, J. V. and Kochar, S. C. (1989). A distribution-free test for the equality of failure rates due to two competing risks. *Comm. Statist. Theory Methods* **18**, 107–120

Baringhaus, L. (1991). Testing for spherical symmetry of a multivariate distribution. *Ann. Statist.* **19**, 899-917.

Baringhaus, L. and Henze, N. (1991). Limit distributions for measures of skewness and kurtosis based on projections. *J. Multiv Anal.* **38**, 51-69.

Barnard, G.A. (1963). Discussion of Professor Bartlett's paper. *J. R. Statist. Soc.* B. **25**, 294.

Bartlett, M.S.(1963). The spectral analysis of point processes (with discussion). *J. R. Stat. Soc.,* B. **25**, 264-296.

Beran, R. (1979). Testing for elliptical symmetry of a multivariate density. *Ann. Statist.* **7**, 150-162.

Beran, R. and Ducharme, G. R. (1991). Asymptotic theory for bootstrap methods in statistics. Université de Montréal, Centre de Recherches Mathmatiques, Montreal, QC.

Beran, R and Srivastava, M. S. (1985). Bootstrap tests and confidence regions for functions of a covariance matrix. *Ann. Statist.* **13**, 95-115.

Besag, J. and Diggle, P. J. (1977). Simple Monte Carlo tests for spatial pattern. *Appl. Statist.* **26**, 327-333.

Bickel, P. (1978). Using residuals robustly I: Tests for heteroscedasticity. *Ann. Statist.* **6**, 266-291.

Block, H. W. and Basu, A. P. (1974). A continuous bivariate exponential distribution. *J. Am. Statist. Assoc.* **69**, 1031-1037.

Blough, D. K. (1989). Multivariate symmetry viaprojection pursuit. *Ann. Inst. Statist. Math.* **41**, 461-475.

Boos, D. D. and Brownie, C. (1989). Bootstrap methods for testing homogeneity of variances. *Technometrics* **31**, 69-82.

Box, G. E. P. (1949). A general distribution theory for a class of likelihood criteria. *Biometrika* **36**, 317-346.

Buckley, M. J. (1991). Detecting a smooth signal: optimality of cusum based on procedures. *Biometrika* **78**, 253-262.

Burke M. D. and Yuen, K. C. (1995). Goodness-of-fit tests for the Cox model via bootstrap method. *J. Statist. Plann. Inf.* **47**, 237-256.

Butler, C. C. (1969). A test for symmetry using the sample distribution function. *Ann. Math. Statist.* **14**, 2209-2210.

Cai, Z., Fan, J. and Li, R. (2000). Efficient estimation and inferences for varying-coefficient models. *J. Amer. Statist. Assoc.* **95**, 888–902.

Cai, Z., Fan, J. and Yao, Q. (2000). Functional-coefficient regression models for nonlinear time series. *Journal of American Statistical Association* **95**, 941-956.

Carroll, R. J. (1982). Adapting for heteroscedasticity in linear models, *Ann. Statist.* **10**, 1224-1233.

Carroll, R. J., and Ruppert, D. (1981). On robust tests for heteroscedasticity. *Ann. Statist.* **9**, 205-209.

Carroll, R. J., and Ruppert, D. (1988). *Transformation and Weighting in Regression.* Chapman and Hall, New York.

Chiang, C. T, Rice, J. A and Wu, C. O. (2001). Smoothing spline estimation for varying coefficient models with repeatedly measured dependent variables. *J. Am. Statist. Ass.* **96**, 605-619.

Cook, R. D. and Weisberg, S. (1983). Diagnostics for heteroscedasticity in regression. *Biometrika* **70**, 1-10.

Cox, D. D., Koh, E., Wahba , G. and Yandell, B. S. (1988). Testing the (parametric) null model hypothesis in (semiparametric) partial and generalized spline models. *Ann. Statist.* **16**, 113-119.

Csörgö, S. , and Heathcote, C.R. (1987). Testing for symmetry. *Biometrika* **74**, 177-184.

Cuzick, J. (1992). Semi-parametric additive regression. *J. Roy. Statist. Soc. Ser. B* **54**, 831-843.

Davidian, M. and Carroll, R. J. (1987). Variance function estimation. *J. Amer. Statist. Assoc.* **82**, 1079-1091.

Davidian, M., and Giltinan, D. M. (1995). *Nonlinear Models for Repeated Measurement Data.* Chapman and Hall, London

Davison, A. C. and Hinkley, D. V. (1997). *Bootstrap Methods and Their Application*. Cambridge University Press, UK.

Dawkins, B. (1989). Multivariate analysis of national track records. *The American Statistician* **43**, 110-112.

Dempster, A. P. (1969). *Elements of Continuous Multivariate Analysis*. Addison-Wesley, USA.

Dette, H. (1999). A consistent test for the functional form of a regression based on a difference of variance estimators. *Ann. Statist.* **27**, 1012-1040.

Dette, H. and Munk, A. (1998). Testing heteroscedasticity in nonparametric regression. *J. R. Statist. Soc. B* **60**, 693-708.

Diblasi, A. and Bowman, A. (1997). Testing for constant variance in a linear model. *Statist. and Probab Letters* **33**, 95-103.

Diggle, P. J., Heagerty, P. J.,Liang K-Y and Zeger, S. L. (2002). *Analysis of Longitudinal Data*. Oxford University Press, Oxford, England.

Diks, C. and Tong, H. (1999). A test for symmetries of multivariate probability distributions. *Biometrika* **86**, 605-14.

Doksum, K. A., Fenstad, G., and Aaberge, R. (1977). Plots and tests for symmetry. *Biometrika* **64**, 473-487.

Dudley, R. M. (1978). Central limit theorems for empirical measures. *Ann. Probab.* **6**, 899-929.

Dunnett, C. W. (1994). Recent results in multiple testing: Several treatments vs. a specified treatment. *Proceedings of the International Conference on Linear Statistical Inference LINSTAT'93, Math. Appl. 306*, 317-346. Kluwer Acad. Publ., Dordrecht, The Netherlands.

Eaton, M. L. and Tyler, D. E. (1991). On Wielandt's inequality and its application to the asymptotic distribution of the eigenvalues of a random symmetric matrix. *Ann. Statist.* **19**, 260-271.

Efron, B. (1979). Bootstrap methods: Another look at the jackknife. *Ann. Statist.* **7**, 1-26.

Efron, B. and Tibshirani, R. (1993). *An Introduction to the Bootstrap*. Chapman and Hall, New York.

Engle, R. F., Granger, C. W. J., Rice, J. and Weiss, A. (1986). Semiparametric estimates of the relation between weather and electricity sales. *J. Amer. Statist. Assoc.* **81**, 310-320.

Engen, S. and Lillegård, M. (1997). Stochastic simulations conditioned on sufficient statistics. *Biometrika* **84**, 235-40.

Eubank, R. L. and Hart, J. D. (1992). Testing goodness-of-fit in regression via order selection criteria. *Ann. Statist.* **20**, 1412-1425.

Eubank, R. L. and Hart, J. D. (1993). Commonality of cusum, von Neumann and smoothing-based goodness-of-fit tests. *Biometrika* **80**, 89-98.

and LaRiccia, V. N. (1993). Testing for no effect in nonparametric regression. *J. Statist. Plann. Inference* **36**, 1-14.

Eubank, R. L. and Thomas, W. (1993). Detecting heteroscedasticity in nonparametric regression. *J. Roy. Statist. Soc. Ser. B* **55**, 145-155.

Fan, J. and Huang, L. (2001). Goodness-of-fit tests for parametric regression models. *J. Amer. Statist. Assoc.* **96**, 310-320

Fan, Y. and Li, Q. (1996). Consistent model specification tests: omitted variables and semiparametric functional forms. *Econometrica* **64**, 865-890.

Fan, J. and Zhang, W.Y. (1999). Statistical estimation in varying coefficient models. *Ann. Statist.* **27**, 1491-1518.

Fan, J. and Zhang, W.Y. (2000). Simultaneous confidence bands and hypothesis testing in varying-coefficient models. *Scandinavian Journal of Statistics*, **27**, 715-731.

Fang, K. T., Kotz, S. and Ng, K. W. (1990). *Symmetric Multivariate and Related Distributions*. Chapman and Hall, London.

Fang, K. T. and Li, R. Z. and Zhu, L. X. (1997). Some probability plots to test spherical and elliptical symmetry. *J. Comp. Graph. Stat.* **6**, 435-450.

Fang, K. T., Zhu, L. X., and Bentler, P. M. (1993). A necessary test of goodness of fit forsphericity. *J. Multiv. Anal.* **44** , 34-55.

Feuerverger, A., and Mureika, R. A. (1977). The empirical characteristic function and is applications. *Ann. Statist.* **5**, 88-97.

Friedman, J. H. (1987). Exploratory projection pursuit. *J. Amer. Statist. Assoc.* **82**, 249-266.

Gaenssler, P. (1983). *Empirical Processes*. Lecture Notes-Monograph series 3 Institute of Mathematical Statistics, Hayward, California.

Ghosh, S., and Ruymgaart, F.H. (1992). Applications of empirical characteristic functions in somemultivariate problems. *Canad. J. Statist.* **20**, 429-440.

Giné, E. and Zinn, J. (1984). On the central limit theorem for empirical processes (with discussion). *Ann. Probab.* **12**,929-998

Giné, E. and Zinn, J. (1990). Bootstrapping general empirical measures. *Ann. Probab.* **18**, 851-869

Good, P. (2000). *Permutation Tests: A practical guide to resampling methods for testing hypothesis*. Second edition, Springer, New York

Gozalo, P. L. and Linton, O. B. (2001). Testing additivity in generalized nonparametric regression models with estimated parameters. *J. Econometrics* **104**, 1-48

Gu, C. (1992). Diagnostics for nonparametric regression models with additive terms. *J. Amer. Statist. Assoc.* **87**, 1051-1058.

Guttman, I. and TIAO, G. C. (1965). The inverted Dirichlet distribution with applications. *J. Am. Statist. Assoc.* **60**, 793-805.

Hall, P. and Titterington, D. M. (1989). The effect of simulation order on level accuracy and power of Monte Carlo tests. *J. R. Statist. Soc.* B **51**, 459-67.

Hall, W.J. and Wellner, J.A. (1984). Mean residual life. In *Proc. Int. Symp. Statistics and Related Topics* eds M. Csörgő, D.A. Dawson, J.N.K. Rao and A.K. Md.E. Saleh, pp. 169-184. Amsterdam: North-Holland.

Härdle, W. (1990). *Applied nonparametric regression*. Cambridge University Press, New York.

Härdle, W. and Mammen, E. (1993). Comparing non-parametric versus parametric regression fits. *Ann. Statist.* **21**, 1926-1947.

Härdle, W., Mammen, E. and Müller, M. (1998). Testing parametric versus semiparametric modeling in generalized linear models. *J. Amer. Statist. Assoc.* **93**, 1461-1474.

Hart, J.D. (1997). *Nonparametric Smoothing and Lack-of-fit Tests*, Springer, New York.

Hastie, T. and Tibshirani, R. (1993). Varying-coefficient models (With discussion). *J. Roy. Statist. Soc. Ser. B* **55**, 757–796.

Heathcote, C. R., Rachev, S.T., and Cheng, B. (1995). Testing multivariate symmetry. *J. Multiv. Anal.* **54**, 91-112.

Henze, N. and Wagner, T. (1997). A new approach to the BHEP tests for multivariate normality. *J. Mult. Anal.* **62**, 1-23.

Hope, A. C. A. (1968). A simplified Monte Carlo test procedure. *J. R. Statist. Soc.* B **30**, 582-98.

Hoeffding, W.(1952). The large-sample power of tests based on permutations of observations. *Ann. Math. Stat.* **23**, 169-192.

Hoel, D. G. (1972). A representation of mortality data by competing risks. *Biometrics* **28**, 475-488.

Hoover, D. R., Rice, J. A, Wu, C. O and Yang, L-P. (1998). Nonparametric smoothing estimates of time-varying coefficient models with longitudinal data. *Biometrika* **85**, 809-822.

Huang, J.Z., Wu, C.O., and Zhou, L. (2002). Varying-coefficient models and basis function approximations for the analysis of repeated measurements. *Biometrika* **89**, 111-128.

Huang, J.Z., Wu, C.O., and Zhou, L. (2004). Polynomial spline estimation and inference for varying coefficient models with longitudinal data. *Statistica Sinica* **14**, 763-788.

Jennrich, R. I. (1969). Asymptotic properties of non-linear least squares estimators. *Ann. Math. Statist.* **40**, 633-643.

Jing, P. and Zhu, L. X.(1996). Some Blum-Kiefer-Rosenblatt type tests for the joint independence of variables. *Comm in Statist.: Theory and Methods* **25**, 2127-2139.

Johnson, R. A. and Wichern, D. W. (1992). *Applied Multivariate Statistical Analysis*. 3rd Ed. Singapore: Prentice Hall, Simon & Schuster Asia.

Kariya, T. and Eaton, M. L. (1982). Robust tests for spherical symmetry. *Ann. Statist.* **1**, 206-215.

Kaslow, R. A, Ostrow, D. G., Detels, R., Phair, J. P., Polk, B. F. and Rinaldo, C.R. (1987). The Multicenter AIDS Cohort Study: rationale, organization and selected characteristics of the participants. *Am. J. Epidem.* **126**, 310-318.

Kim, J. (2000). An order selection criteria for testing goodness of fit. *J. Amer. Statist. Assoc.* **95**, 829-835.

Korin, B. P. (1968). On the distribution of a statistic used for testing a covariance matrix. *Biometrika* **55**, 171-178.

Koul, H. L. (1992). *Weighted Empiricals and Linear Models*. Lecture Notes— Monograph Series, 21. Institute of Mathematical Statistics, Hayward, California.

Lam, K. F. (1997). A class of tests for the equality of k cause-specific hazard rates in a competing risks models. *Biometrika* **85**, 179-188.

Li, K. C. (1991). Sliced inverse regression for dimension reduction (with discussions). *J. Amer. Statist. Assoc.* **85**, 316-342.

Li, R. Z, Fang, K. T. and Zhu, L. X. (1997). Some Q-Q probability plots to test spherical and elliptical symmetry. *J. Comp. Graph. Statist.* **6**, 435-450.

Liang, K.-Y., and Zeger, S. L. (1986). Longitudinal data analysis using generalized linear models. *Biometrika* **73**, 13-21.

Liu, R. Y. (1988). Bootstrap procedures under some non-i.i.d. models. *Ann. Statist.* **16**, 1696-1708.

Maguluri, G. and Zhang, C.H. (1994). Estimation in the mean residual life regression model. *J. R. Statist. Ser.* B **56**, 477-489.

Mammen, E. (1992) *When Does Bootstrap Work? Asymptotic Results and Simulations. Lecture Notes in Statistics 77* Springer, New York.

Mammen, E. and van de Geer, S. (1997). Penalized quasi-likelihood estimation in partial linear models. *Ann. Statist.* **25**, 1014–1035.

Marshall, A. W. and Olkin, I. (1979). *Inequalities: Theory of Majorization and Its Applications.* New York: Academic Press.

Miller, B. M., Runggaldier, W. J. (1997). Kalman filtering for linear systems with coefficients driven by a hidden Markovjump process. *Syst. Control Lett.* **31**, 93-102.

Müller, H. G. (1992). Goodness-of-fit diagnostics for regression models. *Scand. J. Statist.* **19**, 157-172.

Müller, H. G. and Zhao, P. L. (1995). On a semi-parametric variance function model and a test for heteroscedasticity. *Ann. Statist.* **23**, 946-967.

Muirhead, R. J. (1982). *Aspect of Multivariate Statistical Theory.* John Wiley, New York.

Naik, D. N. and Khattree, R.(1996). Revisiting Olympic track records: Some practical considerations in the principal component analysis. *The American Statistician* **50**, 140-144.

Neuhaus, G (1991). Some linear and nonlinear rank tests for competing risks models. *Comm. Statist.: Theory Methods* **20**, 667–701.

Neuhaus, G. and Zhu, L. X. (1998). Permutation tests for reflected symmetry. *J. Multivariate Anal.* **67**, 129-153.

Nolan, D. and Pollard, D. (1987). U-process: Rates of convergence. *Ann. Statist.* **15**, 780-799.

Nolan, D. and Pollard, D. (1988). Functional limit theorems for U-process. *Ann. Probab.* **15**, 1291-1299

Oakes, D. and Dasu, T. (1990). A note on residual life. *Biometrika* **77**, 409-410.

O'Brien, R. D. (1979). A general ANOVA method for robust tests for aditive models for variances. *J. Amer. Statist. Assoc.* **74**, 877-880.

O'Brien, R. D. (1981). A simple test for variance effects in experimental designs. *Psychological Bulletin* **89**, 570-574.

Olkin, I. and Rubin, H. (1964). Multivariate Beta distributions and independence properties of the Wishart distributions. *Ann. Math. Statist.* **35**, 261-69.

Pearson, E. S. and Hartley, H. O. (1972). *Biometrika Tables for Statisticians,* Vol. 2. Cambridge University Press, Cambridge.

Pollard, D. (1984). *Convergence of Stochastic Processes.* Springer-Verlag, New York.

Præstgaard, J. P. (1995). Permutation and bootstrap Kolmogorov-Smirnov test for the equality of two distributions. *Scand. J. Statist.* **22**, 305-322.

Roger, P. Q., Jun, S., and Mari,P. (2001). Efficiency comparison of methods for estimation in longitudinal regression models. *Statist. and Probab. Lett.* **55**, 125-135.

Romano, J. P. (1989). Bootstrap and randomization tests of some nonparametric hypotheses. *Ann. Statist.* **17**, 141-159.

Rothman, E. D., and Woodroofe, M. (1972). A Cramer-von Mises type statistic for testing symmetry. *Ann.Math. Statist.* **43**, 2035-2038.

Roy, S.N. (1953). On a heuristic method of test construction and its use in Multivariate analysis. *Annals of Math. Stat.* **24**, 220-238

Royston, J. P. (1983). Some techniques for assessing multivariate normality based on the Shapiro-Wilk W. *Appl. Statist.* **32**, 121-133.

Schick, A. (1996). Root-n consistent estimation in partly linear regression models. *Statist. Probab. Lett.* **28**, 353-358.

Shorack, G. R. and Wellner, J. A (1986). *Empirical Processes with Applications to Statistics.* Wiley, New York.

Schuster, E. F., and Barker,R. C. (1987). Using the bootstrap in testing symmetry andasymmetry. *Camm. Statist. Simul. Comp.* **16**, 69-84.

Shao, J. and Tu, D. (1995). *The Jackknife and Bootstrap.* Springer-Verlag, New York.

Shorack, G., and Wellner, J. A. (1986). *The Empirical Processes with Applications to Statistics.* Wiley, New York.

Simonoff, J. S. and Tsai, C. L. (1991). Assessing the influence of individual observations on a goodness-of-fit test based on nonparametric regression. *Statist. Prob. Lett.* **12**, 9-17.

Singh, K. (1981). On the asymptotic accuracy of Efron's bootstrap. *Ann. Statist.* **9**, 1187–1195.

Small, N. J. H. (1980). Marginal skewness and kurtosis in testing multivariate normality. *Appl. Statist.* **29**, 85-87.

Speckman, P. (1988). Kernel smoothing in partial linear models. *J. Roy. Statist. Soc. Ser. B* **50**, 413-436.

Spokoiny, V. G.(1996). Adaptive hypotheesis testing using wavelets. *Ann. Statist.* **24**, 2477-2498.

Stone, C. J. (1982). Optimal global rates of convergence for nonparametric regression. *Ann. Statist.* **10**, 1040-1053.

Stute, W. (1997). Non-parametric model checks for regression. *Ann. Statist.* **25**, 613-641.

Stute, W. and Manteiga, G. W. (1995). NN goodness-of-fit tests for linear models. *J. Statist. Plann. Inf.* **53**, 75-92.

Stute, W., Manteiga, G. W. and Quindimil, M. P. (1998). Bootstrap approximations in model checks for regression. *J. Amer. Statist. Asso.* **93**, 141-149.

Stute, W., Thies, G. and Zhu, L. X. (1998). Model checks for regression: An innovation approach. *Ann. Statist.* **26**, 1916-1934.

Stute, W. and Zhu, L. X. (2002). Model Checks For Generalized Linear Models. *Scan. J. Statist.* **29**, 535-546.

Stute, W., and Zhu, L. X. (2005). Nonparametric Checks For Single-Index Models. *Ann. Statist.* **33**, to appear.

Stute, W., Zhu, L.X. and Xu, W. L. (2005). Dimension Reduction Tests for Parametric Regression Models. Submitted for publication.

Su, J.Q. and Wei, L.J. (1991). A lack-of-fit test for the mean function in a generalized linear model. *J. Amer. Statist. Assoc.* **86**, 420-426.

Sun Y. Q., Wu. H. L. (2004). Semiparametric time-varying coefficients regression model for longitudinal data. Unpublished manscript.

Sykes, L. R., Isacks, B. L. and Oliver, J. (1969). Spatial distribution of deep and shallow earthquakes of small magnitudes in the Fiji-Tonga region. *Bull. Seismol. Soc. Am.* **59**, 1093-1113.

van der Vaart, A. W. and Wellner, J. A. (2000). *Weak Convergence and Empirical Processes.* Springer, New York.

Vonesh, E. F. and Chinchilli, V. W. (1997). *Linear and Nonlinear Models for the Analysis of Repeated Measurements.* Marcel Dekker, New York.

Wahba, G. (1984). Cross validation spline methods for the estimation of multivariate functions from data on functionals. *In Statistics: An Appraisal, Proc. 50th Anniversary Conf. Iowa State Statistical Laboratory* (H. A. David and H. T. David, eds) 205-235. Iowa State University Press, Ames.

Ware J. H. (1985). Linear models for the analysis of longitudinal studies.*Amer. Statist.* **39**, 95-101.

Whang, Y. and Andrews, D. W. K. (1993). Tests of specification for parametric and semiparametric models. *J. Econometrics* **57**, 277-318.

Wu, C. F. J. (1986). Jackknife, bootstrap and other re-sampling methods in regression analysis. *Ann. Statist.* **14**, 1261-1295.

Wu, C. O., Chiang, C. T., and Hoover, D.R. (1998), Asymptotic Confidence Regions for Kernel Smoothing of a Varying-Coefficient Model With Longitudinal Data. *J. Amer. Statist. Assoc.* **93** 1388-1402.

Wu, C. O., and Chiang, C. T. (2000). Kernel smoothing on varying coefficient models with longitudinal dependent variable. *Statistic Sinica 10*, 433-456.

Wu, H. and Liang, H. (2004). Backfitting random varying-coefficient models with time-dependent smoothing covariates. *Scan. J. Statist. 31*, 3-19.

Xu, W. L. and Zhu, L. X. (2004). Goodness-of-fit Tests for a Varying-Coefficients Model in Longitudinal Studies. Submitted for publication.

Yatchew, A. J. (1992). Nonparametric regression tests based on least squares. *Econometric Theory* **8**, 435-451.

Yuen K.C. and Burke M.D. (1997). A test of fit for a semiparametric additive risk model. *Biometrika* **84**, 631-639.

Yuen, K. C., Zhu, L. X. and Tang, N.Y. (2003). On the mean residual life regression model. *J. Statist. Plan. Inf.* **113**, 685-698.

Yuen, K. C., Zhu, L. X. and Tang, N. Y. (2001). On the mean residual life regression model. Technical report, Department of Statistics and Actuarial Science,

Zeger, S. L., Liang, K. Y. and Albert, P. S. (1988). Models for longitudinal data: A generalized estimation equation approach. *Biometrics* **44**, 1049-1060.

Zhang, J. and Boos, D.D. (1992). Bootstrap critical values for testing homogeneity of covariance matrices. *J. Amer. Statist. Assoc.* **87**, 425-429.

Zhang, J., and Boos, D.D. (1993). Testing hypothesis about covariance matrices using bootstrap methods. *Comm. in Statist.: Theory and Methods.* **22**, 723-739.

Zhang, J., Pantula, S. G. and Boos, D. D. (1991). Robust methods for testing the pattern of a single covariate matrix. *Biometrika* **78**, 787-795.

Zhu, L.X. (1993). Convergence rates of empirical processes indexed by classes of functions and their applications. *J. Syst. Sci. Math. Sci.* **13**, 33–41 (in Chinese)

Zhu, L. X. and Fang, K. T. (1994). The accurate distribution of Kolmogorov statistic with Bootstrap approximation. *Advanced in Appl. Math.* **15**, 476-489.

Zhu L. X. (2003). Model checking of dimension-reduction type for regression. *Statist. Sinica* **13**, 283-296.

Zhu, L. X. and Cui, H. J. (2005). Tsting the Adequacy for A General Linear Errors-in-varibles Model. *Statistica Sinica* **15**, to appear.

Zhu, L.X. and Fang, K. T. (1994). The accurate distribution of Kolmogorov statistic with Bootstrap approximation. *Advanced in Appl. Math.* **15**, 476-489.

Zhu, L. X. and Fang, K. T. (1996). Asymptotics for kernel estimate of sliced inverse regression. *Ann. Statist.* **14**, 1053-1068.

Zhu, L. X., Fang, K. T. and Bhatti, I. M. (1997). On estimated projection pursuit type Cramer-von Mises statistics. *J. Mult. Anal.* **63**, 1-15.

Zhu, L.X., Fang, K. T., Bhatti, I. M. and Bentler, P. M. (1995). Testing sphericity of a high-dimensional distribution based on bootstrap approximation. *Pakistan J. Statist.* **14**, 49-65.

Zhu, L. X., Fang, K. T. and Li, R. Z. (1997). A new approach for testing symmetry of a high-dimensional distribution. *Bull. Hong Kong Math. Soc.* **1** 35-46.

Zhu, L. X., Fang, K. T., and Zhang, J. T. (1995). A projection NT-type test for spherical symmetry of a multivariate distribution. *Multivariate Statistics and Matrices in Statistics eds, E. -M. Tiit, Kollo, T. and Niemi, H.* TEV & VSP, Holland, 109-122.

Zhu, L. X., Fujikoshi, Y. and Naito, K. (2001). Heteroscedasticity test for regression models. *Science in China, Series A* **44**, 1237-1252.

Zhu, L. X. and Jing, P. (1998). On some tests based on projection pursuit for elliptical symmetry of a high-dimensional distribution. *Chinese Bull. Sci.*, **43**, 450-457.

Zhu, L. X. and Ng, K. W. (2003). Checking the adequacy of a partial linear model. *Statist. Sinica* **13**, 763-781.

Zhu, L. X., Ng, W. and Jing, P. (2002). Resampling methods for homogeneity tests of covariance matrices. *Statist. Sinica* **12**, 769-783.

Zhu, L. X. and Neuhaus, G. (2000). Nonparametric Monte Carlo test for multivariate distributions. *Biometrika* **87**, 919-928.

Zhu, L. X. and Neuhaus, G. (2003). Conditional tests for elliptical symmetry. *J.Multiv. Anal.* **84**, 284-298.

Zhu, L. X. and Zhu, R. Q. (2005). Model Checking for Multivariate Regression Models. Submitted for publication.

Index

adaptive Neyman test, 60
Aki, 4

Baringhaus, 4
Barnard, 1
Bartlett, 1
Bentler, 4
Beran, 2, 4, 84, 155
Besag, 1
Bhatti, 1, 4
Bickel, 103
Boos, 155, 162
Bowman, 103
Box, 155
Burke, 146, 151

Cai, 124
Carroll, 103
Chebychev inequality, 52
Cheng, 4, 29
Chiang, 123, 133
classical bootstrap, 2, 47, 60, 108, 142, 145, 158, 162
conditional Monte Carlo test, 4
Cook, 103
Crämer-von Mises, 8
Cui, 87, 88
Cuzick, 59

Dasu, 141
Davidian, 103
Davison, 2
Dawkins, 94
Dempster, 30

Dette, 59, 103
Diblasi, 103
Diggle, 1
Diks, 4
dimension-reduction type test, 44
Ducharme, 2, 84
Dudley, 6
Dunnett, 157

Efron, 2, 84, 108
Engen, 1
Engle *et al.*, 59
Eubank, 59, 103

Fan, 60, 62, 123, 133
Fang, 1, 2, 4
Friedman, 25
Fujikoshi, 138

Gaenssler, 53, 54, 74
generalized cross validation, 66, 110
Ghosh, 4, 36
Giné, 6, 116, 152, 165
Glivenko-Cantelli theorem, 39
Gonzá lez Manteiga, 3, 8, 48, 59, 104
González Manteiga, 89
Good, 3
Guttman, 12

Härdle, 3, 8, 59, 89, 109
Hölder inequality, 53
Hall, 1
Hart, 59
Heathcote, 4, 29

Henze, 13
heteroscedasticity, 103
Hinkley, 2
Hoeffding, 32
Hoeffding inequality, 53, 75, 78
Hoover, 123
Hope, 1
Huang, 60, 62, 123, 133

Isacks, 67

Jennrich, 126
Jing, 155
Johnson, 84, 88, 94

Kaslow, 133
Kolmogorov, 2, 8
Kotz, 2
Koul, 53

LaRiccia, 59
Li, 26, 124
Liang, 123
likelihood ratio test, 88, 155
Lillegård, 1
Liu, 109
local alternative, 27–29, 64, 107, 110

Müller, 103
Maguluri, 141, 142
Mammen, 3, 8, 59, 109
Mammen , 89
mean residual life, 141
Monte Carlo test
   MCT, 1, 2
multivariate distribution
   elliptical symmetry, 25, 26, 90
   independently decomposable, 4, 5,
      11, 13
   Liouville-Dirichlet, 11, 18
   reflection symmetry, 11, 12, 17, 28
   spherical symmetry, 11, 12, 16, 37
   spherically symmetric, 32
   symmetric scale mixture, 12, 19
multivariate regression, 83
Munk, 103

Naito, 138
Neuhaus, 4, 11, 25
Ng, 2, 60, 155

Nolan, 72, 118

O'Brien, 157
Oakes, 141
Oliver, 67
Olkin, 12

Pantula, 155
partially linear model, 59, 68
permutation test, 3, 5, 160
Pollard, 6, 39, 40, 52, 54, 71, 72, 75, 78,
      116, 118, 121, 150, 154, 165
Presedo Quindimil, 3, 8, 48, 60, 89, 104
Præstgaard, 165, 166

Rachev, 4, 29
random symmetrization, 6
Rice, 123
Romano, 32
Roy, 155
Royston, 20
Rubin, 12
Ruppert, 103
Ruymgaart, 4, 36

Shao, 2
Shorack, 111
Singh, 2
smoothing based test
   CUSUM test, 43
   globally smoothing method, 43, 133
   globally smoothing test, 60
   innovation transformation based test,
      43
   locally smoothing method, 43, 133
   locally smoothing test, 60
   score type test, 8, 43, 84, 92
Speckman, 59
Spokoiny, 60
Srivastava, 155
Stochastic process
   conditional empirical process, 31, 34
   empirical process, 26
   fidis convergence, 39, 77, 121, 138,
      153
   Gaussian process, 27, 28, 46, 71, 76
   residual marked empirical process,
      48, 60, 61
   uniform tightness, 39, 40, 77, 121,
      138, 153

Stone, 69
Stute, 3, 8, 48, 59, 87–89, 104, 125, 128
Sun, 124
Sykes, 67

Tang, 142, 153
Taylor expansion, 37, 151
Thies, 60, 104, 125, 128
Thomas, 103
Tiao, 12
Titterington, 1
Tong, 4
Tu, 2

U-statistic, 71, 72, 75

validity of test
    asymptotic validity, 32, 34, 41
    exact validity, 5, 31
van de Geer, 59
van der Vaart, 6, 7
VC class, 72, 74, 77, 154

Wagner, 13
Wald, 77
Weisberg, 103
Wellner, 6, 7, 111
Wichern, 84, 88, 94
Wild bootstrap
    wild bootstrap, 3, 6, 9, 60, 109
Wilks lambda, 89, 91, 94, 95
Wu, 3, 89, 109
Wu, C., 123, 133

Yang, 123
Yuen, 142, 146, 151, 153

Zhang, 123, 133, 141, 142, 155, 162
Zhao, 103
Zhou, 133
Zhu, 1, 2, 4, 11, 25, 44, 60, 87, 88, 94,
    104, 125, 128, 138, 142, 153, 155
Zinn, 6, 116, 152, 165

Lecture Notes in Statistics

For information about Volumes 1 to 127, please contact Springer-Verlag

128: L. Accardi and C.C. Heyde (Editors), Probability Towards 2000. x, 356 pp., 1998.

129: Wolfgang Härdle, Gerard Kerkyacharian, Dominique Picard, and Alexander Tsybakov, Wavelets, Approximation, and Statistical Applications. xvi, 265 pp., 1998.

130: Bo-Cheng Wei, Exponential Family Nonlinear Models. ix, 240 pp., 1998.

131: Joel L. Horowitz, Semiparametric Methods in Econometrics. ix, 204 pp., 1998.

132: Douglas Nychka, Walter W. Piegorsch, and Lawrence H. Cox (Editors), Case Studies in Environmental Statistics. viii, 200 pp., 1998.

133: Dipak Dey, Peter Müller, and Debajyoti Sinha (Editors), Practical Nonparametric and Semiparametric Bayesian Statistics. xv, 408 pp., 1998.

134: Yu. A. Kutoyants, Statistical Inference For Spatial Poisson Processes. vii, 284 pp., 1998.

135: Christian P. Robert, Discretization and MCMC Convergence Assessment. x, 192 pp., 1998.

136: Gregory C. Reinsel, Raja P. Velu, Multivariate Reduced-Rank Regression. xiii, 272 pp., 1998.

137: V. Seshadri, The Inverse Gaussian Distribution: Statistical Theory and Applications. xii, 360 pp., 1998.

138: Peter Hellekalek and Gerhard Larcher (Editors), Random and Quasi-Random Point Sets. xi, 352 pp., 1998.

139: Roger B. Nelsen, An Introduction to Copulas. xi, 232 pp., 1999.

140: Constantine Gatsonis, Robert E. Kass, Bradley Carlin, Alicia Carriquiry, Andrew Gelman, Isabella Verdinelli, and Mike West (Editors), Case Studies in Bayesian Statistics, Volume IV. xvi, 456 pp., 1999.

141: Peter Müller and Brani Vidakovic (Editors), Bayesian Inference in Wavelet Based Models. xiii, 394 pp., 1999.

142: György Terdik, Bilinear Stochastic Models and Related Problems of Nonlinear Time Series Analysis: A Frequency Domain Approach. xi, 258 pp., 1999.

143: Russell Barton, Graphical Methods for the Design of Experiments. x, 208 pp., 1999.

144: L. Mark Berliner, Douglas Nychka, and Timothy Hoar (Editors), Case Studies in Statistics and the Atmospheric Sciences. x, 208 pp., 2000.

145: James H. Matis and Thomas R. Kiffe, Stochastic Population Models. viii, 220 pp., 2000.

146: Wim Schoutens, Stochastic Processes and Orthogonal Polynomials. xiv, 163 pp., 2000.

147: Jürgen Franke, Wolfgang Härdle, and Gerhard Stahl, Measuring Risk in Complex Stochastic Systems. xvi, 272 pp., 2000.

148: S.E. Ahmed and Nancy Reid, Empirical Bayes and Likelihood Inference. x, 200 pp., 2000.

149: D. Bosq, Linear Processes in Function Spaces: Theory and Applications. xv, 296 pp., 2000.

150: Tadeusz Caliński and Sanpei Kageyama, Block Designs: A Randomization Approach, Volume I: Analysis. ix, 313 pp., 2000.

151: Håkan Andersson and Tom Britton, Stochastic Epidemic Models and Their Statistical Analysis. ix, 152 pp., 2000.

152: David Ríos Insua and Fabrizio Ruggeri, Robust Bayesian Analysis. xiii, 435 pp., 2000.

153: Parimal Mukhopadhyay, Topics in Survey Sampling. x, 303 pp., 2000.

154: Regina Kaiser and Agustín Maravall, Measuring Business Cycles in Economic Time Series. vi, 190 pp., 2000.

155: Leon Willenborg and Ton de Waal, Elements of Statistical Disclosure Control. xvii, 289 pp., 2000.

156: Gordon Willmot and X. Sheldon Lin, Lundberg Approximations for Compound Distributions with Insurance Applications. xi, 272 pp., 2000.

157: Anne Boomsma, Marijtje A.J. van Duijn, and Tom A.B. Snijders (Editors), Essays on Item Response Theory. xv, 448 pp., 2000.

158: Dominique Ladiray and Benoît Quenneville, Seasonal Adjustment with the X-11 Method. xxii, 220 pp., 2001.

159: Marc Moore (Editor), Spatial Statistics: Methodological Aspects and Some Applications. xvi, 282 pp., 2001.

160: Tomasz Rychlik, Projecting Statistical Functionals. viii, 184 pp., 2001.

161: Maarten Jansen, Noise Reduction by Wavelet Thresholding. xxii, 224 pp., 2001.

162: Constantine Gatsonis, Bradley Carlin, Alicia Carriquiry, Andrew Gelman, Robert E. Kass Isabella Verdinelli, and Mike West (Editors), Case Studies in Bayesian Statistics, Volume V. xiv, 448 pp., 2001.

163: Erkki P. Liski, Nripes K. Mandal, Kirti R. Shah, and Bikas K. Sinha, Topics in Optimal Design. xii, 164 pp., 2002.

164: Peter Goos, The Optimal Design of Blocked and Split-Plot Experiments. xiv, 244 pp., 2002.

165: Karl Mosler, Multivariate Dispersion, Central Regions and Depth: The Lift Zonoid Approach. xii, 280 pp., 2002.

166: Hira L. Koul, Weighted Empirical Processes in Dynamic Nonlinear Models, Second Edition. xiii, 425 pp., 2002.

167: Constantine Gatsonis, Alicia Carriquiry, Andrew Gelman, David Higdon, Robert E. Kass, Donna Pauler, and Isabella Verdinelli (Editors), Case Studies in Bayesian Statistics, Volume VI. xiv, 376 pp., 2002.

168: Susanne Rässler, Statistical Matching: A Frequentist Theory, Practical Applications and Alternative Bayesian Approaches. xviii, 238 pp., 2002.

169: Yu. I. Ingster and Irina A. Suslina, Nonparametric Goodness-of-Fit Testing Under Gaussian Models. xiv, 453 pp., 2003.

170: Tadeusz Caliński and Sanpei Kageyama, Block Designs: A Randomization Approach, Volume II: Design. xii, 351 pp., 2003.

171: D.D. Denison, M.H. Hansen, C.C. Holmes, B. Mallick, B. Yu (Editors), Nonlinear Estimation and Classification. x, 474 pp., 2002.

172: Sneh Gulati, William J. Padgett, Parametric and Nonparametric Inference from Record-Breaking Data. ix, 112 pp., 2002.

173: Jesper Møller (Editor), Spatial Statistics and Computational Methods. xi, 214 pp., 2002.

174: Yasuko Chikuse, Statistics on Special Manifolds. xi, 418 pp., 2002.

175: Jürgen Gross, Linear Regression. xiv, 394 pp., 2003.

176: Zehua Chen, Zhidong Bai, Bimal K. Sinha, Ranked Set Sampling: Theory and Applications. xii, 224 pp., 2003

177: Caitlin Buck and Andrew Millard (Editors), Tools for Constructing Chronologies: Crossing Disciplinary Boundaries, xvi, 263 pp., 2004

178: Gauri Sankar Datta and Rahul Mukerjee , Probability Matching Priors: Higher Order Asymptotics, x, 144 pp., 2004

179: D.Y. Lin and P.J. Heagerty , Proceedings of the Second Seattle Symposium in Biostatistics: Analysis of Correlated Data, vii, 336 pp., 2004

180: Yanhong Wu, Inference for Change-Point and Post-Change Means After a CUSUM Test, xiv, 176 pp., 2004

181: Daniel Straumann, Estimation in Conditionally Heteroscedastic Time Series Models, x, 250 pp., 2004

182: Lixing Zhu, Nonparametric Monte Carlo Tests and their Applications, xi, 189 pp., 2005

 Springer
the language of science

springeronline.com

Estimation in Continually Heteroscedastic Time Series Models

D. Straumann

This monograph concentrates on mathematical statistical problems associated with fitting conditionally heteroscedastic time series models to data. This includes the classical statistical issues of consistency and limiting distribution of estimators. Particular attention is addressed to (quasi) maximum likelihood estimation and misspecified models, along with phenomena due to heavy-tailed innovations. The used methods are based on techniques applied to the analysis of stochastic recurrence equations.

2004. 228 p. (Lecture Notes in Statistics) Softcover ISBN 3-540-21135-7

Inference for Change Point and Post Change Means After a CUSUM Test

Y. Wu

This monograph is the first to systematically study the bias of estimators and construction of corrected confidence intervals for change-point and post-change parameters after a change is detected by using a CUSUM procedure. Researchers in change-point problems and sequential analysis, time series and dynamic systems, and statistical quality control will find that the methods and techniques are mostly new and can be extended to more general dynamic models where the structural and distributional parameters are monitored. Practitioners, who are interested in applications to quality control, dynamic systems, financial markets, clinical trials and other areas, will benefit from case studies based on data sets from river flow, accident interval, stock prices, and global warming.

2004. 176 p. (Lecture Notes in Statistics) Softcover ISBN 0-387-22927-2

Easy Ways to Order▶ Call: Toll-Free 1-800-SPRINGER • E-mail: orders-ny@springer.sbm.com • Write: Springer, Dept. S8113, PO Box 2485, Secaucus, NJ 07096-2485 • Visit: Your local scientific bookstore or urge your librarian to order.